Ophthalmology:

A Brief Review for Nurses, Medical Students, and Ophthalmic Technicians

Justin E. Anderson, MD

APOLLO™
AUDIOBOOKS

NEW YORK, NEW YORK

Find this and other titles available in

Digital Audio, Text Download, and Print at

www.ApolloAudiobooks.com

ISBN-13: 978-0-6151-9996-2

Request for permissions to reproduce any portion of this book should be addressed to
Apollo Audiobooks, LLC
ATTN: Permissions
3204 30th Street
Lubbock, Texas 79410

Apollo Audiobooks is a registered trademark of Apollo Audiobooks, LLC
New York, New York
www.ApolloAudiobooks.com

Cover art by Apollo Audiobooks, LLC

Dedication

This book is dedicated to my wife, Laura D. Anderson,
without whom this work would not have been possible.

Special thanks to Claire Fraser, MD, PhD,
for her useful input into this work.

- ## INTRODUCTION

- ### For Medical Students, Nurses and Technicians

This "Brief Review" was in response to a request from nurses studying for the Certification Examination for Ophthalmic Registered Nurses. The information contained herein can serve as a primer for all members of the ophthalmic healthcare team including nurses, ophthalmic technicians, scrub nurses, operating room technicians, ophthalmic billing and coding specialists, and medical students starting a one or four week elective.

Learning ophthalmology, like many professions, begins with learning the language of ophthalmology. Therefore, much of this text is laid out in a simple format with a disease and its definition and possible treatments. The overall goal of the book is to familiarize the entire ophthalmic healthcare team with the language and practice of ophthalmology in order that the physicians, nurses, technicians, and coders all have a basic grasp of the healthcare environment in which they work.

- ### For Nurses Studying for the CEORN

According to the 2007 NCBORN Handbook for candidates, The Certification Examination for Ophthalmic Registered Nurses is a computer-based examination that contains a maximum of 250 multiple-choice questions. Testing subjects include: Ocular Conditions, Pharmacology, Nursing Assessment of the Ophthalmic Patient, Ophthalmic Nursing Interventions and Patient Education, and Professional Issues.

This "Brief Review" is intended to supplement more comprehensive books and materials, you should use to study for the Certification Examination for Ophthalmic Registered Nurses. The material contained within this review is intended for TEST TAKING PURPOSES ONLY, and is NOT intended to be applied to direct patient care. Consultation and study from comprehensive ophthalmology and nursing texts is necessary for any patient care interaction or healthcare decision making.

- ### For All Other Readers

Remember, this text is intended to be used as a primer or a brief review of ophthalmology. The information contained herein is for EDUCATIONAL AND TEST-TAKING PURPOSES ONLY, and should NOT be used to guide any clinical decision making or patient care interactions. This text is ONLY an abbreviated dictionary of terms and conditions. Consultation and study from comprehensive ophthalmology texts is necessary for any patient care interaction or healthcare decision making.

For Medical Students, Nurses and Technicians

For Nurses Studying for the CGRN

For All Other Readers

• TABLE OF CONTENTS

• SECTION 1: ANATOMY & EMBRYOLOGY

Let's begin with a review of basic ocular anatomy and physiology

• Eyeball Anatomy

Cornea- the clear, transparent window in the front of the eye.
Sclera- white part of the eye that is continuous with the Cornea
Limbus- junction between Cornea and Sclera
Conjunctiva- skin that covers the Sclera
Tenon's Capsule- fibrous tissue that is located between the Conjunctiva and Sclera
Iris- pigmented eye muscle behind the cornea with an opening for light (called the Pupil)
Trabecular Meshwork- at the Cornea/Iris junction, it drains Aqueous Humor from the eye to the Canal of Schlemm
Ciliary Body- pigmented eye muscle behind the Iris that creates Aqueous Humor and changes the shape of the Lens
Lens- focuses light on the Retina (remember, the Cornea and Lens are both necessary to focus light on the Retina. The Cornea is actually more powerful than the Lens)
Zonules- ligaments that extend from the Ciliary Body to the Lens to hold it in place
Anterior Chamber- Aqueous-filled space between the Cornea and Iris
Posterior Chamber- Aqueous-filled space between Iris and Lens
Vitreous Cavity- Vitreous-filled space between the Lens and Retina
Vitreous Humor- a jelly-like substance that becomes more liquid with age
Aqueous Humor- a liquid, plasma like substance that flows from the Posterior Chamber to the Anterior Chamber
Retina- innermost layer of the eye that contains photoreceptor and senses light
Choroid- middle layer of the eye that provides some (1/3) of the blood supply to the Retina via the Choriocapillaris
Uvea – pigmented structures of the eye including Iris, Choroid, and Ciliary Body
Sclera- outermost layer of the eye (see above)
Optic Nerve- sensory nerve that carries messages from the Retina to the Brain
Central Retinal Artery and Vein- pass through the Optic Nerve to supply most (2/3) of the blood supply to the Retina
Fovea- the area of sharp, distinct vision in the retina

• Adnexal Anatomy

Palpebral Fissure- opening between the eyelids
Canthus- the medial and lateral corners of the eyelids
Bulbar Conjunctiva- transparent skin that covers the sclera
Palpebral Conjunctiva- transparent skin that covers the eyelids
Fornix (pl. Fornices)- location where the conjunctiva folds over transitioning from Bulbar to Palpebral Conjunctiva
Tarsus- fibrous tissue that gives structure to the upper and lower lids
Meibomian Glands- sebaceous glands that are found in the Tarsus and open at the eyelid margin
Punctum (pl. Puncta)- the holes in the upper and lower lids medially that drain tears

into the Nasolacrimal System

Lacrimal Glands- large glands under the upper lids temporally used for reflex tearing

Nasolacrimal System- Puncta, Upper and Lower Canaliculi, Common Canaliculus, Lacrimal Sac, and Nasolacrimal Duct draining into the nose

Extraocular Muscles- six muscles that move the eyeball

Rectus Muscles- four eyeball muscles that insert at a straight angle onto the eyeball

Oblique Muscles- two eyeball muscles that insert at an oblique angle onto the eyeball

Levator Palpebrae Superioris- muscle that elevates the upper lid

Muller's Muscle- second muscle that elevates the upper lid (under sympathetic control)

Cranial Nerves- sensory and motor nerves to the face (there are 12 cranial nerves)

Cranial Nerve (CN) 3- opens the eye (Levator Palpebrae) and controls ALL of the Extraocular Muscles EXCEPT Superior Oblique and Lateral Rectus

CN 4- controls the Superior Oblique

CN 6- controls the Lateral Rectus

CN 7- closes the eye by controlling the orbicularis oculi muscle

Ptosis- drooping of the upper eyelid

- ## Embryology

Hyaloid Artery- extends from the Optic Nerve to the posterior portion of the lens capsule. Persistence of the Hyaloid Artery can cause serious vision loss (PHPV: Persistent Hyperplastic Primary Vitreous).

Mittendorf Dot- small opacity on the posterior lens capsule, a remnant of the Hyaloid Artery, NOT visually significant

Bergmeister's Papilla- small extension of vascular tissue extending from the Optic Nerve into the vitreous, also a remnant of the Hyaloid Artery, NOT visually significant

Retinal Vasculature- not complete until the 9th month of gestation, putting premature infants at risk for Retinopathy of Prematurity

Macula and Fovea- not completely developed at birth, continue to mature during the first months of life

• SECTION 2: THE BASICS

• General Terms and Conditions

Accommodation- focusing the eye at near
Diplopia- double vision
Amblyopia- poor vision that cannot be corrected, due to poor visual development in childhood for a variety of reasons (Cataract, Ptosis, Strabismus, or Refractive Error)
Amblyopia Treatment- includes correcting the underlying disease if possible (cataract extraction) and patching of the better-seeing eye
Refraction- correcting the eye's focusing power by using external lenses (eyeglasses)
Myopia- nearsightedness, the pt CAN see better at NEAR (near vision is perfectly clear)
Hyperopia- farsightedness, the pt CAN see better at DISTANCE (however, neither near nor distance vision are perfectly clear)
Presbyopia- the loss of Accommodation (can no longer focus at near), begins around age 40 to 45 and progresses to age 65
Refractive Error- can be corrected with lenses (i.e. myopia)
Non-Refractive Error- cannot be corrected with lenses (i.e. cataract or retinal detachment)
Pinhole Acuity- helps tell the difference between Refractive and Non-Refractive Error. If the vision is better with Pinhole, there is likely a Refractive Error that can be corrected with lenses.
Visual Acuity Testing- should be performed at both near and far to ensure there is no refractive problem with either near or distance vision (Remember, test each eye separately)
Cycloplegic Refraction- testing the Refractive Error of a patient after the instillation of Cycloplegic drops that paralyze Accommodation. This is important for children, Hyperopes, and any patient with Strabismus.
Cyclopegia- often accomplished in the office with the instillation of Cyclopentolate drops with or without Tropicamide, then waiting 30 minutes for the drops to take effect. Atropine drops can also be used at home nightly for three days to ensure complete Cycloplegia has been attained.

Assessing vision in children and infants:
For Infants- covering the amblyopic eye (the eye the child is not using) causes no protest or crying. Covering the better-seeing eye causes the child to cry and protest.
Light Reflex- a pen light or muscle light can be used to look for Strabismus in infants. Hold the light in front of the child and have the child fixate on the light. Check to see that the light reflex in the Cornea is a mirror image on the left and right cornea. If the light reflex is not symmetric, the child may have Strabismus

Color Vision Testing- tested for each eye separately using Ishihara plates, is documented as the number of plates seen over the number of plates shown. If the patient can only read the numbers shown in 7 of the 12 plates in the right eye, but reads all 12 of 12 numbers in the left eye, the color vision is documented as 7/12 OD and 12/12 OS.
Color Deficiency- is the inability to distinguish certain colors (more prominent in boys/men)

Spectacles- lenses that sit in front of the eye to correct Refractive Error
Contact Lenses- lenses that sit on the cornea to correct Refractive Error (used in Keratoconus, High Hyperopia or High Myopia, and for cosmetic reasons)
Refractive Surgery- correcting Refractive Error through eye surgery (two common surgeries involving laser are LASIK and PRK)
LASIK- a flap is cut in the cornea with a keratome, the laser is applied, the flap is replaced (the epithelium is maintained intact on the flap so recovery time is shorter)
PRK- the epithelium is removed with an instrument, the laser is applied, the epithelium must then re-grow to cover the cornea (longer recovery time)

- ### Pediatric and Strabismus Terms and Conditions

Orthophoria- straight eyes, the eye muscles are keeping the eyes aligned
Strabimus (-tropia)- misalignment of the eyes
Strabismus in adults- if it is of new onset, often causes Diplopia (double vision) in adults
Strabismus in children- often does NOT cause Diplopia because the child will Suppress (turn off the input) from one eye. Suppression can lead to Amblyopia
Esotropia- eyes are misaligned inward
Exotropia- eyes are misaligned outward
Hypertropia or Hypotropia- the eyes are misaligned upward or downward
Phoria- any type of a Phoria (i.e. esophoria) is a tendency of the eye to turn, but the eyes are NOT misaligned
Nystagmus- rapid, involuntary eye movements
Pseudo-Strabismus- the appearance of strabismus when the eyes are actually straight (i.e. a child with a large nasal bridge)

- ### Oculoplastics and Orbit Terms and Conditions

Blepharospasm- squeezing of the eyelids
Blepharitis- inflammation of the eyelid margin
Ectropion- eyelid turned out, leads to tearing and dry eye
Entropion- eyelid turned in, leads to Trichiasis and Corneal Abrasion or Ulceration
Trichiasis- the eyelashes touching the eyeball
Hordeolum- ACUTE inflammation of the eyelid glands leading to a red, painful swelling
Chalazion- CHRONIC inflammation of the eyelid glands, causes swelling that is often less red and less painful
Dacryoadenitis- lacrimal GLAND inflammation or infection (at the upper outer eyelid)
Dacryocystitis- lacrimal SAC infection (at the medial eyelid by the nose, pus can often be found at the puncta)
Epiphora- tearing
Exophthalmos- (also called proptosis) the eye sticking out from the orbit
Enophthalmos- the eye sinking into the orbit
Enucleation- Complete removal of the eye
Evisceration- Removal of the eye contents while leaving the sclera intact
Exenteration - Removal of both 1. the eye itself and 2. the orbital contents including fat, extraocular muscles, and other tissues

- ## Cornea and Conjunctiva Terms and Conditions

Corneal Abrasion- a scratch in the corneal epithelium, stains positive the fluorescein dye
Conjunctivitis- inflammation of the conjunctiva (can be viral, bacterial, or allergic)
Pterygium- a growth of conjunctival skin over the cornea
Keratitis- inflammation or infection of the cornea
Keratoconus- cone-shape to the cornea

- ## Uveitis Terms and Conditions

Hypopyon- a layer of white cells seen as a white line layered behind the inferior Cornea
Hyphema- a layer of red blood cells seen as a red line layered behind the inferior Cornea
Uveitis- inflammation of a portion of the Uvea (iris, ciliary body, choroid (all pigmented structures of the eye are Uveal Structures))
Sympathetic Ophthalmia- After a pt suffers injury to one eye, Uveitis can occur in the normal, non-injured eye called Sympathetic Ophthalmia
Endophthalmitis- infection or inflammation within the eyeball itself

- ## Cataract and Lens Terms and Conditions

Cataract- an opacity in the lens
Aphakia- the lens has been removed surgically and not replaced
Pseudophakia- the lens has been removed surgically and replaced with an artificial lens implant

- ## Glaucoma Terms and Conditions

Glaucoma- a disease of pressure within the eye that causes damage to the Optic Nerve
Optic Cup- center area of the Optic Nerve that becomes progressively larger with Glaucoma
Trabeculectomy- Glaucoma surgery that filters Aqueous Humor from the Anterior Chamber to the sub-conjunctival space in order to lower intraocular pressure
Tube Shunt Implant- Glaucoma surgery that places a silicone tube from the Anterior Chamber or Vitreous Chamber into the sub-conjunctival space to lower intraocular pressure
ALT/SLT- laser procedure the applies laser to the trabecular meshwork to lower intraocular pressure
Iridotomy/Iridectomy- surgical formation of a hole in the iris by laser or open surgery that allows fluid to flow freely from the Posterior Chamber to the Anterior Chamber

- ## Retina and Vitreous Terms and Conditions

Vitrectomy- surgical removal of the vitreous
Vitreous or Aqueous Tap- using a needle to remove a sample of Vitreous or

Aqueous, often done for endophthalmitis, and may be followed by injection or antibiotics, "Tap & Inject"

Phacoemulsification- the use of ultrasound to break-up and remove a cataract

- ## **Pharmacology Terms**

Cycloplegics- do two things. 1. paralyze accommodation and 2. dilate the pupil (aka Cyclopentolate)
Mydriatics- only do one thing. 1. dilate the pupil
Miotics- constrict the pupil

• SECTION 3: OPHTHALMIC NURSING BASICS

• The Procedures

Hand Washing- the hands must be washed between every patient contact, contact on different body sites of the same patient, after touching contaminated items, and before and after donning gloves

FOR ALL PROCEDURES- Step one is WASH YOUR HANDS.

Lid Hygiene- often used for Blepharitis or lid crusting. Use a cotton applicator or warm towel with a small amount of baby shampoo to scrub the lashes of the closed eye. Then rise and cleanse throughly with water to remove all soap.

Eye Cleansing- often used for crusting. Use a separate, moist cotton ball for each eye, remove all exudate

Eye Irrigation- often used for chemical burns. Check the pH before beginning irrigation, lie the patient supine, place a drape and kidney basin below the patient's head to catch run-off, use a sterile balanced salt solution or normal saline to irrigate the eye thoroughly while holding the eyelids open. Pay special attention to irrigating the superior and inferior Fornices which may harbor particles and chemicals. The irrigation must be COPIOUS, utilizing at least one liter of fluid before re-checking the pH with pH paper. Supplies include one liter of normal saline, IV tubing, drapes, kidney basin, and topical anesthetic (only as prescribed by the physician). A lid speculum may be helpful and used if a second person is not available to assist in opening the eyelids. Take care to direct the stream of fluid away from the uninvolved eye. Alkali burns, in general, cause more damage than acid burns.

Eversion of the Lids- grasp the lashes of the upper lid and pull it forward, away from the eyeball. Use a cotton-tipped applicator to then push down on the lid while simultaneously pushing back on the lashes with the other hand.

Double-Lid Eversion- requires a second instrument such as a lid retractor. This allows access to and irrigation of the superior fornix.

Drop Instillation- wash hands, pull lower lid down at bony orbit, instruct pt to look up, place one drop in the inferior fornix, do NOT touch the eye or lashes with the bottle (if you do, it is contaminated and must be thrown away), have the pt close gently, and dab the lateral canthus with a tissue

Ointment Instillation- wash hands, pull lower lid down at bony orbit, instruct pt to look up, place 1/2 inch of ointment in the inferior fornix, again do NOT touch the eye or lashes with the bottle, have the pt close gently, and dab the lateral canthus with a tissue

Self-Drop Instillation- instruct the patient to wash his hands, lie in the supine position (or seated if the patient can adequately extend the neck), place bottle in hand opposite the eye to receive the drop (left hand for right eye), place the bottle on the nose bridge for stability, pull down the eyelid at the bony orbit, look up, instill one drop and gently close the eyes. After drops have been instilled in both eyes, Punctal Occlusion by placing the forefingers of each hand at the medial canthus should be performed for two minutes. Alternately, the patient can be instructed to keep the eyes closed without blinking for two minutes. Both eyelid closure and Punctal Occlusion help prevent the eye drop from being immediately drained into the nasolacrimal system.

Subconjunctival and Intraocular Injections- used for longer term administration of a medication in the setting of eye inflammation, infection, macular degeneration, or other disorders. Wash hands and don gloves, instill topical anesthetic drop, draw up desired medication in a tuberculin syringe and place a #30G needle on the syringe. Next a lens speculum should be used to prevent any eyelash contact or contamination, and one drop of topical ocular betadine eye prep instilled, the patient is instructed to look in a certain direction, the medication is injected by the physician, the lid speculum removed and a second drop of betadine eye prep or antibiotic drop is instilled.

Cold Compresses- used to decrease pain and swelling. Place a single gauze pad over the closed eyelid and use a plastic bag or glove filled with crushed ice over the gauze.

Warm Compresses- often used for Chalazia. Place a clean towel under warm water to completely soak it, wring out the excess water, place the warm towel over the closed eyelid for 15 minutes, re-warm and repeat as necessary.

Soft Patch and Shield- often used after cataract or other intraocular surgery. Place antibiotic ointment in the eye as directed by the physician, secure a single, soft eye patch over the closed eyelids with 2 pieces of one-inch tape diagonally across the patch, then secure an eye shield over the soft eye patch with one piece of one-inch tape placed diagonally across the patch.

Eye Shield- often used in cases of ruptured globe or suspected ruptured globe or other severe eye injury to prevent both the staff and the patient from rubbing or touching the eye. Place the hard eye shield over the bony orbital rim and tape into place with a single piece of diagonally placed one-inch tape.

Patch Taping- in order to avoid patient discomfort, the tape should be placed diagonally from the corner of the jaw bone (mandible) to the brow above the opposite eye. This allows the patient to eat without discomfort and prevents the patch from coming loose from jaw movement. The hair should be avoided and the mouth should not be distorted by the tape.

Pressure Patch- often used for corneal abrasion, bleeding, or wound leak. Instill antibiotic ointment as instructed by the physician, fold a soft eye patch in half and place over the closed eyelid, place a second eye patch over the first and tape into place diagonally as above. Then place several pieces of tape across the eye patch using gentle pressure with each pass. The tape across the patch should pass medially and laterally enough to cover the entire eye pad, while the upper and lower ends of the tape must come together at the brow and jaw bone.

Conjunctival Cultures and Smears- used in the setting of bacterial conjunctivitis or dacryocystitis to identify the pathologic organism. Instill topical anesthetic, use either a sterile cotton-tip applicator (moistened with sterile saline or BSS) or a platinum spatula (cleaned with alcohol, passed through flame, and cooled to room temperature), have the patient look up, pull lower lid down at bony orbit, gently sweep the inferior fornix for pus and exudate, immediately plate the swab or spatula in a "C" shape onto agar plates as provided by your institution. Additionally, the specimen may be smeared onto a glass slide for special stains, and a cotton applicator may be deposited directly into thioglycolate broth for additional cultures (break the cotton-tip applicator off inside the broth tube to ensure that the portion contaminated by your fingers does not become a portion of the culture specimen).

Guiding the Blind- Have the pt take your right elbow with their left hand and walk just beside and behind you, warn them of steps and narrow doorways, guide them to a seat by placing your hand and theirs upon the seat cushion or seat back.

The Preoperative Period- instruct the patient regarding proper drop instillation and the number and frequency of drops that will be required both before and after the surgery. Instruct the patient not to eat or drink ANYTHING for 8 hours prior to surgery.

Remove any jewelry, dentures, and/or contact lenses. Encourage the patient to void. Properly identify the patient by at least two means prior to any procedure, and confirm the correct site (correct eye) for the procedure with the patient, the physician, the consent, and by visualizing the correct surgical site as marked on the patient.

• SECTION 4: OCULAR PHARMACOLOGY

• Anesthetics

Topical Anesthetics- Include tetracaine, proparacaine, and benoxinate with fluorescein. Start working in 15 seconds and last about 15 minutes. Repeated instillation is toxic to the corneal epithelium and these agents should never be given to the patient to take home.

• Dyes

Fluorescein- used topically for corneal abrasion and applanation tonometry, and intravenously for fluorescein angiography (stains corneal stroma and floats in the tear film). Approximately 10% of patients will have nausea and vomiting with IV fluorescein, this is a common reaction, but not a true allergy. Anaphylactoid or Anaphylactic reactions are less common, but possible with IV fluorescein.
Rose Bengal- use for the diagnosis of dry eye and superior limbic keratitis, stains devitalized epithelial cells and mucous.

• Antibiotics

Aminoglycosides (Gentamycin and Tobramycin)- good for gram negative organisms and Staph aureus, used for serious gram negative infections, Tobramycin is available as a Fortified Drop from a compounding pharmacy
Bacitracin (ointment)- good for gram positive organisms
Bacitracin/Polymyxin B- good for gram positive and gram negative organisms
Chloramphenicol- good for gram positive and gram negative organisms, Aplastic Anemia is a serious reported complication
Erythromycin- good for gram positive organisms and Chlamydia
Fluoroquinolones (Ciprofloxacin, Gatifloxacin, Levofloxacin, Moxifloxacin, Ofloxacin)- in general, good for both gram positive and gram negative organisms, however, some gram positive organisms such as Staph aureus are becoming resistant
Neomycin/Bacitracin/Polymyxin B (ointment)- Neomycin is good for gram positive and negative organisms. The combination is very broad spectrum. Pts using Neomycin develop contact allergies in 10% of cases.
Neomycin/Gramicidin/Polymyxin B (drops)- good for gram positive and gram negative organisms.
Sulfacetamide- good for gram positive and gram negative organisms (except for the gram negative Pseudamonas aeruginosa)
Tetracycline- good for some gram positive organisms (causes tooth staining)
Trimethoprim/Polymyxin B- good for gram positive and gram negative organisms (except for the gram negative Pseudamonas aeruginosa)
Vancomycin- good for almost all gram positive infections (even those Staph resistant to Fluoroquinolones and Penicillin and Cephalosporins), used for serious gram positive infections, only available as a Fortified drop from a compounding pharmacy

Antibiotic Take-Home Points
Fortified Vancomycin with Fortified Tobramycin- used for serious ocular surface infections such as a corneal ulcer
Bacitracin and Erythromycin- only cover gram positive organisms
Chloramphenicol- can cause Aplastic Anemia
Neomycin- has a high incidence of allergy (10%)

- ## Antivirals

Trifluridine- a drop used for herpes simplex keratitis
Other Antivirals (Famciclovir, Foscarnet, Ganciclovir, Idoxuridine, Vidarabine, Zidovudine)- avail as oral, IV, or for injection, used for a variety of viruses including herpes simple, herpes zoster, CMV, HIV, and Epstein-Barr virus

Antiviral Take-Home Point
Trifluridine- burns and is toxic to the epithelium, should be used on a short-term basis and rapidly tapered over several days.

- ## Antifungals

Natamycin- a drop used for fungal corneal ulcers (the only commercially available antifungal drop)
Amphotericin B- used for serious fungal infections of the eye, often in combination with Natamycin, available as a Fortified drop from a compounding pharmacy (or IV or for injection)
Voraconazole- use for serious fungal infections, may be compounded into a fortified drop or used orally
Other Antifungals (Clotrimazole, Fluconazole, Ketoconazole, Miconazol, Nystatin)- available as oral, IV or topical (not ophthalmic preparations), may be use in combination with other antifungals

Antifungal Take-Home Point
Natamycin- the only available commercial antifungal drop, any others must be compounded by a pharmacy

- ## Corticosteroids

General- used to decrease inflammation within the eye (uveitis) and outside the eye (blepharitis or conjunctivitis). All steroids carry some risk of elevating the intraocular pressure and causing steroid-induced glaucoma as well as worsening infectious keratitis (fungal, bacterial or viral), and causing a specific kind of cataract (posterior subcapsular cataract). The risk of these complications is dependent upon the corticosteroid's potency, it's ability to enter the eye through the cornea, and duration of use.
Dexamethasone- the most potent of the corticosteroids, highest risk of steroid-induced glaucoma

Prednisolone- the second-most potent of the corticosteroids, slightly lower risk of glaucoma
Fluorometholone- a less potent steroid, lower risk of glaucoma
Loteprednol-the least potent of the steroids with lowest risk of steroid-induced glaucoma

Corticosteroids Take Home Points
All corticosteroids- carry some risk of glaucoma, cataract, and worsening infection
Dexamethasone- carries the greatest risk and is the most potent
Loteprednol- carries the lowest risk and is the least potent

- ## Nonsteroidal Anti-Inflammatories (NSAIDS)

NSAIDS (Flurbiprofen, Ketorolac, Diclofenac)- reduce inflammation and pain by inhibiting the cyclo-oxygenase enzyme (COX enzyme), do NOT cause glaucoma or cataract, CAN cause keratitis and stromal melting of the cornea

- ## Other Anti-Inflammatories

Cyclosporin- an immunomodulator used for dry eye to increase tear production and decrease inflammation

- ## Mydriatics (sympathomimetics)

Phenylephrine- Causes pupil dilation only, may increase blood pressure leading to stroke, angina, or heart attack

- ## Cycloplegics (parasympatholytics)

Cause 1. Pupil Dilation and 2. Ciliary Muscle (causing loss of accommodation)
Tropicamide- shortest acting (5 hours)
Cyclopentolate-longer (18 hours), used in clinic for cyclopegic refraction, can cause neurotoxicity in children (especially infants)
Homatropine- longer (2.5 days)
Scopolamine- longer (5.5 days), can cause dizziness and disorientation
Atropine- longest (1.5 weeks), can cause redness, flushing and fever (especially in children)

Cycloplegic Take-Home Points
Tropicamide is the shortest acting and Atropine is the longest. Remember the order Tropicamide, Cyclopentolate, Homatropine, Scopolamine, then Atropine.

- ## Anti-Glaucoma Medications

(to lower the intraocular pressure (all either decrease aqueous humor production or increase aqueous humor outflow)

Cholinergic Agonists- increase outflow (Miotics- cause pupil constriction, all may cause brow ache and decreased vision)
Pilocarpine, Carbachol, Echothiophate Iodide- Pilocarpine is the most frequently used, Echothiophate has severe cholinergic side effects including salivation, diarrhea, and vomiting, Echothiophate is an irreversible cholinesterase inhibitor and contraindicated with Succinylcholine Induction for Anesthesia

Adrenergic Agonists- decrease production (Alpha Agonists- can cause cystoid macular edema in aphakic patients, tachycardia and anxiety)
Epinephrine, Dipivefrin, Apraclonidine, Brimonidine- Brimonidine commonly used for long-term glaucoma treatment but hypersensitivity may occur, Apraclonidine for pre- and post-laser IOP control, Dipivefrin is a pro-drug

Adrenergic Antagonists- decrease production (Beta Antagonists- can slow heart rate, worsen congestive heart failure, and induce asthma attack)
Timolol, Carteolol, Metipranolol, Levobunolol, Betaxolol- all can worsen asthma and heat failure, but Betaxolol (a Beta-1 Selective Beta Blocker) should have less potential for these side effects

Carbonic Anhydrase Inhibitors- decrease production (contain SULFA and can cause Sulfa allergy, may cause tingling of lips, toes, and metallic taste to soda, occasionally may cause Aplastic Anemia)
Acetazolamide, Methazolamide, Dichlorphenamide, Dorzolamide (drop), Dorzolamide/Timolol (combo drop), Brinzolamide- Systemic side effects are generally highest with Acetazolamide pills and lowest with Brinzolamide, 10% of pts using Dorzolamide develop allergy

Prostaglandin Analogues- increase outflow through uveoscleral tract (can cause increased lash growth, skin and iris pigmentation, red eye and cystoid macular edema)
Bimatoprost, Latanoprost, Travoprost, Unoprostone- a common first-line glaucoma medication in the clinic, must warn pts of red eye, lash growth, and skin and iris color change

Hyperosmotic Agents- mechanism may be increase outflow by making blood plasma hyperosmotic or by shrinking the vitreous humor (can cause hyperglycemia, brain shrinkage and hemorrhage, headache, congestive heart failure and kidney damage)
Mannitol, Glycerin, Urea- used in urgent situations of acute glaucoma where topical and carbonic anhydrase inhibitors are ineffective, short-term use, last approximately 5 hours

Anti-Glaucoma Take-Home Points
Adrenergic Antagonists (called Beta Blockers or Beta Antagonists): can cause asthma attack or exacerbate congestive heart failure
Sulfa Allergy: the carbonic anhydrase inhibitors, such as acetazolamide (diamox pills), and dorzolamide (trusopt drops) are made with Sulfa and can lead to sulfa allergy

Adrenergic Agonists: (such as brimonidine (Alphagan)) have a high incidence of follicular conjunctivitis, a hypersensitivity reaction

- ## "Red-Eye" Drops

Vasoconstrictors (naphazoline, tetrahydrozoline, phenylephrine)- in common OTC drops for "red eye"
Antihistamine (pheniramine, antazoline, levocabastine, emedastine)- in common OTC drops for allergy
Combinations (naphazoline/pheniramine/antazoline)- for "red eye" and allergy
Mast Cell Stabilizers (Cromolyn, Ketotifen, Lodoxamide)- prescription for the prevention of allergy/vernal conjunctivitis
Combination Antihistamine/Mast Cell Stabilizers (Azelastine, Ketotifen, Nedocromil, Olopatadine)- for more acute allergy treatment and prevention

"Red Eye" Drops Take-Home Points
Vasoconstrictors: chronic use leads to rebound red eye when the drops are stopped
Antihistamines and Mast Cell Stabilizers: chronic use usually does not lead to side effects or rebound red eye

- ## Artificial Tears and Ointments
(to lubricate the eye)

Artificial Tears- many brands, most contain the preservative benzalkonium chloride which can cause irritation of allergy
Preservative Free Tears- do not contain the preservative, can be administered more frequently, cause less irritation or allergy
Viscous Tears- contain agents to maintain the tear longer in the eye
Ointments- last the longest in the eye, but cause significant blurry vision

- ## Hypertonic Drops
(to dehydrate the cornea, used in pseudophakic bullous keratopathy or recurrent corneal erosion)
Sodium Chloride Drops and Ointments

SECTION 5: NURSING ASSESSMENT OF THE OPHTHALMIC PATIENT

• History

Components of History- correctly identify the patient. Gather Chief Complaint, History of the Present Illness, Past Ocular History, Past Medical History, Current Medications and Drops, Allergies, Social History and Family History

History of the Present Illness- ask timing, severity, influences, constancy, side, and review previous documentation

Describe vision problems- near, far, color, blindness, floaters

Describe pain- sharp, dull headache, photophobia, pressure, burning

Describe symptoms- dryness, tearing, discharge

Describe appearance- ptosis, proptosis, redness

Trauma- must be well-documented, time, place and date, exact incident, any safety glasses, prior emergency treatment, high-speed or low-speed injury

• Vision Assessment

Visual Acuity- uses Snellen eye chart, tests central vision, does not have to be measured at 20 feet IF the chart is calibrated for the correct distance,

Visual Acuity Testing- chart well-lit, pt at 20 feet, test one eye at a time, test with and without glasses (then test vision with a pinhole (with glasses if the pt has them))

Low Vision Acuity Testing- if the pt cannot see the big "E", walk the pt toward the chart (or the chart toward the pt) and document the number of feet at which they see the big "E," if they see it at 6 ft, their vision is 6/400; if they cannot see the "E" at all, document the pts ability to count fingers at a certain distance (counts fingers at 2 ft), or to see hand motion, or to perceive a bright light (muscle light)

Near Vision Testing- use a Jaeger Card (near card) and record the pts vision as J1 at measured at the appropriated distance printed on the card (every card may be different), do this with and without spectacles as well

Children's Vision Testing- Six Weeks (tracks light), Three months (Fix and Follow), Six Months (Central Steady Maintained, maintains attention on an object), One Year (CSM, Allen characters), Three Years (Allen and Tumbling "E"), Five Years (Allen and Snellen Letters)

Amsler Grid- tests central vision for scotoma (missing spot in vision) or metamorphopsia (bent lines in vision), test one eye with glasses, stare at central spot in card, not any missing or bent lines (used for Macular Degeneration)

Confrontation Visual Fields- test one eye for the ability of the patient to count fingers in all four quadrants of vision (Superotemporal and Superonasal, Inferotemporal, and Inferonasal)

Tangent Screen- Tests the central 25 degrees of visual field by moving a white test object against a black, felt background

Perimetry (Visual Field Testing)- used in glaucoma, neurologic, and other optic nerve disease

Humphrey VF Testing- automated static perimetry where a computer shows a series of light flashes

Goldmann VF testing- manual dynamic perimetry where a technician moves a series of lights across the patient's visual field

Color Vision- affected in optic nerve diseases and congenital color deficiency, screening for color deficiency done with Ishihara Plates, classifying color deficiency done with Farnsworth Panel of colors mounted in caps

• Refraction

Basics- the basic element of a Refraction are Sphere and Cylinder

Sphere (or Spherical Power)- what makes a person Myopic (near-sighted), or Hyperopic (far-sighted)

Cylinder (or Astigmatic Power) - what gives a person astigmatism, this occurs when the optical system (cornea and lens) are not completely spherical (shaped like an oval or egg, rather than a perfect sphere or basketball)

Autorefractor- machine that estimates the patient's refraction in Sphere and Cylinder

Keratometer- Measures the shape of the cornea, how spherical is the anterior cornea's shape (many autorefractors have an autokeratometer)

Cross Cylinder- lens used measure the axis and strength of a patient's cylinder

Diopter- the unit of power of both spherical and cylindrical lenses

Phoropter- machine with many lenses used for refraction

Trial Frame- frame in which different lenses can be inserted for refraction

• Contact Lens Refraction and Instruction

Contact Lens Refraction- useful in a patient with keratoconus, a hard contact lens is placed in each eye, vision is tested in each eye, and the patient is refracted with a phoropter while wearing hard contact lenses (rigid gas permeable (RGP) lenses)

Rigid Gas Permeable Lenses- wash hands, wash lens with tap water and lens soap, rinse off all soap with tap water, moisten lens with RGP lens solution, pull down lower lid at orbit, insert lens, remove by displacing downward or with a suction device

Soft Contact Lenses- wash hands, open a new lens, NEVER wash with tap water, rinse with soft contact lens solution, pull down lower lid at orbit, insert lens on cornea, remove by pt looking up, pulling lens down onto conjunctiva, and pinching lens between fingers

- ## SECTION 6: PROCEDURES AND PRACTICES IN THE OFFICE AND THE OR

- ### The Basics

Preoperative Interview- identify the patient's name, birth date, operative site, and signed consent
Surgical Scrub- may differ between institutions, recommended 5 minute initial scrub with iodine or chlorhexadine scrub, and 3 to 5 minute subsequent scrubs, alcohol-based waterless scrubs have been shown to be as effective as antiseptic soap scrubs
Ocular anesthesia- topical anesthetic drop administered (proparacaine), IV sedation administered, retrobulbar block administered (0.75% Marcaine and 2% Lidocaine (with or without epinephrine for hemostasis)), lid block administered
Surgical Prep- after block, confirm site, administer proparacaine, done sterile gloves, administer ophthalmic 5% povidone iodine drops using cotton-tipped applicators to open the lids, scrub the lids with a cotton applicator dipped in ocular 5% povidone iodine, paint face, lids and lashes with betadine three times (begin at eyes and circle toward nose and hair), remove excess betadine with moistened cotton, drape pt with sterile sheets under the head and around the hair with towel clip, drape the remainder of the patient, place a sterile plastic window drape over the operative eye, cut open with Stevens scissors

- ### Laser Surgery

Laser Principles: a laser can be used in the office or in the OR for a variety of treatments, it is important to tell the patient what to expect in terms of visual outcome because most patients think "laser" means "LASIK" and they assume that any kind of laser treatment is to give them better, clearer vision (which is not the case in most laser procedures), by the way, cataracts are NOT removed with laser, they are usually removed with phacoemulsification (ultrasound waves)
Panretinal Photocoagulation (PRP): the laser is shot onto the retina to kill the retinal cells in that area, this is used to decrease the amount of factors (such as VEGF) released by sick, ischemic retinal cells, this is used to treat neovascularization that can bee seen in proliferative diabetic retinopathy or after a central retinal vein occlusion, PRP does not improve the pts vision, but helps stop neovascularization which can eventually lead to hemorrhage, retinal detachment and total blindness
Focal Laser Photocoagulation: the laser is again shot onto the retina, but this time the laser spots are smaller and focused only on small, abnormal vessels in the retina called microaneuryms, these microaneuryms are leaking fluid into the retina causing Macular Edema, the laser closes the microaneuryms, stops the leakage, and the edema improves (hopefully the patient's vision will improve as well over time)
LASIK and PRK: the laser is focused on the cornea and shapes the cornea to focus light better on the retina, the pts vision should improve (that's the goal of this type of laser)
Laser Iridotomy: the laser is focused on the iris to create a hole in the iris that connects the posterior chamber to the anterior chamber, this allows the fluid (aqueous) to freely flow from the posterior to the anterior chamber, this is performed for narrow

angle glaucoma, and does not change vision

YAG Laser Capsulotomy: after cataract surgery, the intraocular lens is placed inside the capsular bag, this bag may opacify with time causing decreased vision, the laser is focused on the posterior portion of the capsular bag and used to open a hole in the opacified bag, this should improve the patient's vision

Laser Trabeculoplasty (ALT/SLT): the laser is focused on the trabecular meshwork and burns are made, this helps increase aqueous flow from the anterior chamber to the trabecular meshwork, this is used in open angel glaucoma and should not change vision

Laser Iridoplasty (PICP): laser applied to the peripheral iris 360 degrees to pull the iris away from the angle and relieve the narrow angle, used in patients with persistent narrow angles because of anatomy (called plateau iris)

Laser Types: Argon and Diode lasers use focused light to create burns, YAG lasers use pulsed light energy to create small explosions (plasma formation), for this reason, YAG lasers are used to make holes in things (YAG laser capsulotomy and Laser Iridotomy)

- **In-Office Procedures**

Punctal Plugs: used for dry eye, place in the inferior puncta, may be collagen (temporary), or silicone (permanent), risks include the plug falling out, or pyogenic granuloma formation (inflammatory reaction around the plug)

Punctal Dilation and Irrigation: used to test for nasolacrimal duct obstruction in adults, the puncta are dilated with a punctal dilator, then saline is irrigated through the canaliculus on an irrigation needle, if there is no flow (or a lot of resistance) there is a nasolacrimal duct obstruction, if the patient tastes the saline in their throat, there is not an obstruction (called a positive taste test)

Shirmer Test: used to test for dry eye, special filter paper strips are placed into the inferior fornix for 5 minutes (with or without topical anesthesia), and the amount of tears are measured

Pars Plana: the pars plana is the location in the retina behind the ciliary body and in front of the anterior edge of the retina, this is the preferred location for intraocular injections and vitrectomy surgery because the eye can be entered here with minimal risk of bleeding, causing retinal holes or detachment, or hitting the lens or ciliary body

Intraocular Injections: used to inject medicine for ARMD, or infections or to inject air or gas for superior retinal holes with retinal detachment, the injections are made through the pars plana, it is important to use topical anesthetic, an anesthetic cotton applicator applied to the injection site, ocular 5% povidone iodine before injection, a wire speculum to keep the lashes out of the field, gloves, and postoperative antibiotics - all to prevent endophthalmitis, risks include endophthalmitis, retinal, lens, or ciliary body damage

Tap and Inject: used for endophthalmitis diagnosis and treatment, a needle is placed into the anterior chamber and aqueous is removed and sent for culture, then antibiotics are injected through the pars plana to treat the infection, risks are the same as above

Anterior Chamber Tap and Inject: used for anterior chamber or corneal infections (serious bacterial or fungal ulcers), the a needle is used to remove aqueous from the anterior chamber and sent for culture, then antibiotics are injected into the anterior chamber, risks include damage to the lens and spread of infection from the outside of the eye to the inside of the eye

Sub-Tenon's Injection: used for steroids in uveitis and antibiotics in infection, sterile technique must be observed as for intraocular injections, but the medication is injected under the conjunctiva and tenon's capsule, risks with steroid injections include a high incidence of steroid induced glaucoma (could occur in 40 to 60 percent of patients)

- **Cornea Surgery**

Penetrating Keratoplasty (PK): used for corneal opacities, haze, edema with pain, or keratoconus, the center of the pt's cornea is removed, then similar sized transplant is cut from a cadaveric donor and sewn into place, pts often require long-term steroid drops to prevent rejection, sutures periodically break and must be pulled, and rigid gas permeable lenses may be necessary for the best vision possible after transplant, risks include infection, glaucoma, and ruptured globe with eye trauma

Pterygium Excision: used for pterygia that are symptomatic or vision threatening, the pterygium is excised off the cornea and part of the base is removed from the conjunctiva, then, either a conjunctival autograft or an amniotic membrane graft is sewn or glued with fibrin glue into the conjunctival defect to prevent recurrence, mitomycin c or beta irradiation may be used to prevent recurrence, risks include pterygium recurrence or scleral melting with mitomycin c

LASIK and PRK: used to correct refractive error, in LASIK a flap is cut with a keratome, in PRK the epithelium is debrided but no flap is cut, in both procedures a laser is applied to correct refractive error, risks include infection and late flap dislocation in LASIK and postoperative haze in PRK

- **Lens and Cataract Surgery**

"Extracapsular" Surgery: the capuslar bag is maintained intact so that an intraocular lens (PMMA or Acrylic or Silicone) can be injected into the capsular bag at the end of surgery, both phacoemulsification and extracapsular cataract extraction are types of "Extracapuslar" Surgery, any cataract surgery carries the risk of expulsive hemorrhage, endophthalmitis, retinal detachment, corneal endothelial damage and the loss of the cornea, vision, or the eye

Phacoemulsification: used for most cataracts today, best for lenses that are not hard (not a dense, mature cataract), a small corneal incision is made, and ultrasound is used to break up the lens and remove it with a phacoemulsification hand piece, quicker recovery time and shorter visual rehabilitation time (most recent surgery)

Extracapsular Cataract Extraction: used for mature cataracts (very brown, hard cataracts) where phacoemulsification may cause damage to the corneal endothelium, a larger incision is made, and the entire lens is expressed out of the capsular bag and out of the eye, then incision is sutured, longer recovery and visual rehabilitation time (more recent surgery)

Intracapsular Cataract Extraction: used in the setting of a traumatized lens or cataract with loose zonules, usually done via the pars plana approach today with a fragmatome (ultrasound instrument) and vitrector to remove the vitreous and remaining zonules, a large incision can be utilized or a smaller pars plana vitrectomy incisions, the entire lens and capsular bag are removed together, no lens can be placed in the capsular bag because it is removed, place an anterior chamber lens or leave the patient aphakic (without a lens) and correct with spectacles or contact lens, or suture an IOL to the iris or pars plana

- **Retina and Vitreous Surgery**

Scleral Buckle: used for retinal detachments due to a well localized tear, good for inferior detachments (because the air used in a vitrectomy surgery will not support a hole and keep it closed superiorly), the conjunctiva is opened and a piece of silicone is placed in the area of the retinal tear, a silicone encircling band is placed around the eye to hold the sponge in place over the break, the fluid under the retina may then be drained via a hole in the sclera (sclerotomy), and laser or cryotherapy placed around the tear, risks include anterior segment ischemia (cutting off the blood supply to the anterior eye) from a tight buckle, or strabismus due to muscle manipulation

Pars Plana Vitrectomy: used for superior tears with retinal detachment, tractional detachments due to proliferative diabetic retinopathy or other proliferative disease, endophthalmitis, and many other retinal conditions, three holes are placed in the pars plana, one for an infusion of BSS, the other for a light pipe (to illuminate the eye), and the third for a vitrector or other second instrument, the vitreous is removed with the vitrector, then several other maneuvers can be added such as endolaser (to laser around holes or for PRP), membrane peel (to remove proliferative neovascular membranes causing retinal detachments), air fluid exchange (the BSS is removed from the eye and air is injected to flatten the area of detached retina), intraocular gas (air, SF6 or C3F8 is injected and left in the eye to hold down the retina and hole after the surgery-lasts days to 8 weeks after surgery), or silicone oil injection (silicone oil is injected and left in the eye up to a year or longer to provide permanent support to the retina-used for recurrent detachments, diffuse retinal disease or infection, and detachments due to PVR), risks include hemorrhage, endophthalmitis, recurrent detachment and lens damage

Proliferative Vitreoretinopathy (PVR): the number one cause for failed retinal detachment surgery, there is a proliferation of glial and retinal tissue that pulls on the retina and causes it to re-detach

- **Uveitis Surgery**

Ganciclovir Implant: used for CMV retinitis, an incision is made in the pars plana and an implant containing ganciclovir is placed in the eye and sewn to the sclera, risks include endophtlamitis, retinal detachment and lens damage

Steroid Implant: investigational use for chronic uveitis, similar to a ganciclovir implant but contains steroids

- **Glaucoma Surgery**

Laser Iridotomy: used for narrow angle glaucoma, the laser is focused on the iris to create a hole in the iris that connects the posterior chamber to the anterior chamber, this allows the fluid (aqueous) to freely flow from the posterior to the anterior chamber, risks include inflammation, post-laser IOP spike, and bleeding (hyphema)

Laser Trabeculoplasty (ALT/SLT): used for open-angle glaucoma (best for pigmentary and pseudoexfoliation glaucomas), laser applied to the trabecular meshwork to increase outflow of aqueous through the trabecular meshwork into the bloodstream, risks include post-laser pressure spike and uveitis

Laser Iridoplasty (PICP): laser applied to the peripheral iris 360 degrees to pull the iris away from the angle and relieve the narrow angle, used in patients with persistent narrow angles because of anatomy (called plateau iris)

Trabeculectomy: used for many types of glaucoma (bad for neovascular glaucoma because of the high risk of bleeding) a trap-door type opening is cut in the sclera to allow a new escape route for aqueous from the anterior chamber to the subconjunctival space, the conjunctiva is then sewn over the trap door, may use mitomycin c or 5-FU at the site of the bleb (area where the aqueous collects under the conjunctiva) to prevent scarring and bleb failure, risks include hemorrhage, hypotony (a persistently low pressure), failed bleb, blebitis (bleb infection), or endophthalmitis

Tube Shunt Placement (Seton Valve Placement): used for many types of glaucoma (often when other surgeries have failed or in neovascular glaucoma), a silicone tube is placed into the anterior chamber or the posterior chamber that leads to the subconjunctival space, the tube is the new escape route for aqueous, risks are similar to trabeculectomy and also include strabismus due to muscle manipulation

Goniotomy/Trabeculotomy: used for infants and children with congenital glaucoma and abnormal angle structures, either the trabecular meshwork is cut from within the eye using a knife (goniotomy), or from the outside of the eye with a special instrument or suture in the canal of schlemm (trabeculotomy), risks include failure, hemorrhage, inflammation and endophthalmitis

- ## Neuroophthalmology Surgery

Optic Nerve Sheath Fenestration: used for pseudotumor cerebri to lower the CSF pressure surrounding the optic nerve, the medial rectus muscle is taken off the eye and the eye is torted laterally, the optic nerve sheath is located and slits are cut into the sheath and CSF is allowed to drain freely, the medial rectus is then replaced, risks include damage to the optic nerve, meningitis (infection of the CSF), and failure

Lumbar Puncture: a frequent in hospital procedure used for pseudotumor cerebri and diagnosis of MS and other CNS diseases, a needle is placed through the back into the spinal cord and fluid is drained out until the CSF pressure is normalized, the fluid may be sent for studies, risks include nerve damage, spinal infection, pain at the procedure site and severe headache

- ## Oculoplastics: The Eyelids and Orbit Surgery

Enucleation: used for blind, painful, non-infected eye, the entire eye is removed including sclera, usually the muscles are sewn to a donor sclera with an implant, performed for blind, painful eye, risks include hemorrhage and infection

Evisceration: used for blind, painful, infected eye (such as endophthalmitis), the contents of the eye are removed by opening the cornea and spooning out the tissue, leaving the sclera in place, the muscles are already attached to the sclera, so they are not moved, an advantage compared to enucleation is that the orbit is not entered so infection should not spread to the orbit, a disadvantage is that the result is not as cosmetically appealing, has decreased risk of infection

Exenteration: used for malignant orbital tumors involving the eye and soft tissues, the entire eye and orbital contents are removed, this is the least cosmetically appealing of the three but at times medically necessary

Canthotomy and Cantholysis: used for emergency decompression of the orbit in the case of retrobulbar hemorrhage, the lateral canthus is incised (canthotomy) and the lateral canthal tendon of the lower lid is cut (cantholysis)

Lateral Tarsal Strip: used for lower lid laxity that leads to entropion or ectropion, a canthotomy and cantholysis are performed, then the tarsus of the lower lid is sewn to the periosteum in a position higher and tighter than the original insertion, risks include bleeding, infection and lower lid asymmetry

Ptosis Repair: used for ptosis, may include advancement of the levator or a mullerectomy (both to pull the upper lid up higher), risks include bleeding, infection and lid asymmetry

Levator Recession: used for lagophthalmos with lid retraction, the levator is moved away from the tarsus to allow the lid to fall further down, risks are the same

Tarsorrhaphy: used for exposure keratitis or severe ulcerative keratitis with risk of perforation, the lateral one-third of the lid is sewn shut (permanently or temporarily) to protect the eye, risks are the same

Dacryocystorhinostomy: used for nasolacrimal duct obstruction (NLDO) in adults, a portion of the bone in the lacrimal fossa is removed and new connection is made between the lacrimal sac and the nose, a silicone tube is placed as a stent in the new system for several months, risks include severe bleeding (can be life-threatening) and infection

Probing and Irrigation: used for congenital NLDO in infants, (best done around 12 months of age for best success rate), a probe is passed through the superior and inferior puncta to the nasal cavity to open the membrane over the valve of Hasner, then saline is irrigated, risks include failure and infection

Blepharoplasty: used for blepharoptosis, excess skin is removed from the upper lid, risks include infection and bleeding

Orbital Decompression: used for thyroid orbitopathy with exposure and proptosis, the medial orbital wall is removed and the lateral orbital wall is thinned out, this allows for more room for the orbital contents so the eye can fall backwards more into the orbit, risks include bleeding, infection, retrobulbar hematoma, and optic nerve damage

Anterior Orbitotomy: used for tumor excision, the orbit it entered only anterior to the orbital septum

Posterior Orbitotomy: used for deeper tumor excision, the orbit is entered posterior to the orbital septum, risks include retrobulbar hematoma and optic nerve damage

- ## Pediatric Ophthalmology and Strabismus Surgery

Eye Muscle Surgery: used for a variety of -tropias to restore normal anatomy, relieve diplopia, restore fusion, rectify a head posture, or promote more normal visual development, may be any combination of muscles depending on the pts deviation, the muscles may be resected (strengthened and made shorter to pull more), recessed (weakened and moved backward to pull less), or transposed (move to a totally different position to pull in a different direction), risks include failed surgery, late recurrence of strabismus, muscle fibrosis, and infection

Recess/Resect Procedure (R&R): used for esotropia or exotropia, is surgery on 2 muscles in only one eye, one muscle is resected (strengthened) and its antagonist is recessed (weakened) to pull the eye in a certain direction, for example, if the eye is turned inward (esotropia), the medial rectus would be recessed and the lateral rectus (its antagonist) would be resected, all of this to pull the eye outward, an advantage of R&R is that the other eye (which has not been operated on) can be easily operated on in the future for any residual deviation, a disadvantage is limitation of motility in the operated eye

Bilateral Surgery: used for esotropia or exotropia, is surgery on 1 muscle in each eye, can be bilateral medial rectus recession or resection, or bilateral lateral rectus recession or resection, an advantage is that the pt does not have motility restriction, a disadvantage is that either eye will be more difficult to operate on in the future because both eyes have scarring

Three Muscle Surgery (or More): used for very large deviations where one or two muscle surgery will not relieve the deviation, risks of three muscle surgery include anterior segment ischemia (because the blood supply to the anterior portion of the eye is carried in the extraocular muscles, when these muscles are cut and moved, the blood supply is cut as well

Transposition Surgery: used for a paretic muscle that no longer has any force or pull, one or two muscles will be transposed (or partially transposed to the site of the weak muscle, for example, if a pt has esotropia because of lateral rectus palsy (the lateral rectus can no longer pull the eye out), all or half of the superior and inferior rectus muscles can be moved beside the lateral rectus to pull the eye out to a more straight position

Vertical Muscle Surgery: used for vertical deviations, one muscle is moved to relieve the deviation

Oblique Muscle Surgery: used alone or in combination with horizontal muscle surgery in pts with oblique muscle overaction that causes an A or V pattern

• SECTION 7: CORNEA AND EXTERNAL DISEASE: A BRIEF REVIEW

• The Basics

Tear Film: covers cornea, made of 3 layers (mucin, aqueous, lipid)
Cornea: made of 3 basic layers (epithelium, stroma, endothelium
Conjunctiva: epithelium covering Tenon's capsule and sclera

• Lid Conditions

These conditions cause redness, irritation and/or itching of the Eyelids.

Blepharitis: inflammation of lids, may be bacterial (staphylococcal) or seborrheic (dandruff), or rosacea (skin condition), treated with warm compresses, antibiotics, and lid scrubs
Hordeolum: acute inflammation of eyelid meibomian gland, treated with warm compresses, and antibiotic drops with or without steroid drops
Contact Dermatitis: skin allergy of the eyelids to some medication or cosmetic, treated by stopping the offending agent, may give topical steroid cream
Rosacea: skin condition that causes a red face, can be associated with blepharitis, can be treated with topical metronidazole or systemic tetracycline

• Conjunctival Conditions

These conditions cause redness, irritation, itching and/or pain of the conjunctiva (red eyes)

Conjunctivitis: inflammation of the conjunctiva for a variety of reasons
Phlyctenulosis: a conjunctival reaction to bacteria (staphylococcal or Tuberculosis), causes red conjunctival nodule, treated with antibiotic and steroid drops
Allergic Conjunctivitis: an acute reaction that causes red, itchy eye and swelling, treated with cold compresses, antihistamines, and mast cell stabilizers
Atopic Keratoconjunctivitis: inflammation of the cornea and conjunctiva associated with eczema and asthma, treated with antihistamines and topical or oral steroids
Vernal (Seasonal) Conjunctivitis: seasonal, recurring conjunctivitis, also associated with eczema and asthma, can cause a shield ulcer (superior corneal ulcer) if severe, treated with antihistamines, topical or oral steroids, and topical cylcosporin drops
Follicles: collections of lymphocytes on the conjunctiva that occur with viral or Chlamydial conjunctivitis or reaction to chronic drop use
Papillae: edema of the conjunctiva that occurs with bacterial or allergic conjunctivitis
Neonatal Conjunctivitis: may be due to silver nitrate drops or infection (Gonococcus, Chlamydia, other bacteria, or virus), the treatment consists of topical and systemic antibiotics for Gonococcus and Chlamydia to prevent systemic complications and blindness
Viral Conjunctivitis: due to several viruses, all are highly contagious and pts should be advised to minimize contact with others and wash hands thoroughly and often

Epidemic Keratoconjunctivitis (EKC): Severe conjunctivitis caused by Adenovirus, can cause membranes (that must be stripped off in the office), decreased vision, corneal infiltrates (opacities), and severe pain, treated with cold compresses and topical steroid drops

Inclusion Conjunctivitis: chronic Chlamydial conjunctivitis in adults, infection with chlamydia, treated with oral azithromycin to prevent permanent scarring that can cause blindness

Trachoma: conjunctival and lid scarring that is the result of chronic Chlamydial conjunctivitis, treated with azithromycin if caught early, treated with lubrication and surgery if caught late (after scarring has occurred)

Ocular Cicatricial Pemphigoid: slow, chronic autoimmune conjunctival scarring that causes red, dry eye, treated with systemic immunosuppressives and lubrication

Stevens-Johnson Syndrome: acute eruption of the mucosal body surfaces including conjunctiva, can be related to medications (especially sulfa) or viral infections, acutely treated with hospital admission and chronically treated with lubrication and surgery for dry eye and scarring

Floppy Eyelid Syndrome: lid eversion in sleep that causes chronic conjunctivitis, occurs in overweight pts, treated with eye shield, lubrication, and surgery if severe

Giant Papillary Conjunctivitis (GPC): chronic conjunctivitis due to contact lenses or ocular prosthesis, treated by decreasing contact/prosthetic use and with steroid drops

Keratoconjunctivitis Sicca (dry eye): caused by many conditions, commonly thyroid, Sjogren syndrome, or poor tear film, treated with lubricants, cyclosporin drops, and eyelid surgery if lid position is the cause

Vitamin A Deficiency: dry eye due to keratinization of the conjunctiva and loss of goblet cells, treated with vitamin A and lubricants

- **Congenital Conditions**

These conditions are present from birth and can cause vision loss, amblyopia, or glaucoma

Anterior Segment Dysgenesis: several syndromes that can cause poor vision from birth leading to amblyopia and permanent vision loss

Peters' Anomaly: problem with lens and cornea, pts have leukoma (white spot where the cornea touches the lens and/or iris), treated with cataract surgery and/or corneal transplant in infancy, pts need amblyopia treatment including patching

Axenfeld's Syndrome and Rieger's Anomaly: conditions with malformed peripheral iris and trabecular meshwork, predispose patients to glaucoma, treated with glaucoma surgery in childhood (goniotomy or trabeculotomy) to open the trabecular meshwork

Aniridia: absence of the iris, can develop nystagmus and poor vision, children need amblyopia treatment and kidney ultrasounds (check for Wilms tumor of the kidney)

Iridocorneal Endothelial Syndromes (ICE Syndromes): endothelial and iris abnormalities that can cause glaucoma, treat glaucoma with drops or surgery

Birth Trauma: cause of cloudy cornea at birth, due to forceps delivery, later causes horizontal endothelial lines called striae (Haab's striae)

Microcornea: small cornea (<10mm in children, <9mm in newborns) associated with Weill-Marchesani, Ehlers-Danlos, and Reiger's Syndromes

Megalocornea: large cornea (>13mm) associated with Marfan Syndrome, craniosynostosis, and ichthyosis

Sclerocornea: sclera in the place where corneal tissue should be, appears as a very white eye, cannot see the iris, treatment is penetrating keratoplasty

- ## Corneal Noninfectious Conditions

These conditions cause eye pain (the cornea is very sensitive), redness and blurry vision

Keratitis: inflammation of the cornea due to many reasons, causes painful red eye, loss of the epithelium, and possibly corneal ulcer (loss of epithelium and corneal stroma)
Staphylococcal Marginal Keratitis: immune reaction to Staph aureus on the lids, causes red eye, treated with steroid/antibiotic combination drops
Recurrent Erosion: corneal abrasion that occurs repetitively without trauma, caused by old trauma or corneal disease, treated with lubrication, sodium chloride drops, bandage contact lens, or surgery (anterior stromal micropuncture) to get the skin to adhere to the corneal stroma better
Medicamentosa Keratitis: corneal ulcer due to medication toxicity, common with topical anesthetic abuse (never let the pt walk out with an anesthetic bottle), or tobramycin or trifluridine drops, treat by stopping the offending drop and use topical lubricants
Exposure Keratitis: keratitis due to poor lid closure, treated with lubrication and eyelid surgery
Neurotrophic Keratitis: keratitis due to loss of the nerve sensation in the cornea, may be due to surgery or Herpes virus infection, treated with lubrication, bandage contact lens or surgery (temporary or permanent tarsorrhaphy)

- ## Neoplastic (Tumorous) Conditions

These conditions are growths on the conjunctiva or lids that can be malignant or benign

Conjunctival Papilloma: (a wart on the conjunctiva) a benign growth caused by papillomavirus, may be excised
Squamous Cell Carcinoma of the Conjunctiva: a malignant skin growth of the conjunctiva due to UV exposure, treated with surgery to excise the cancer, cryotherapy, and topical chemotherapy drops (mitomycin)
Nevus: (a mole on the conjunctiva) a benign proliferation of melanocytes, treated with serial photographs and observation
Primary Acquired Melanosis: brownish color of the conjunctiva, a premalignant condition, predisposes pts to malignant melanoma (especially if the patient is fair skinned)
Melanoma: malignant proliferation of melanocytes, treated with lesion excision and enucleation is possible if the eye is invaded by the tumor, these pts require systemic workup for metastases
Pyogenic Granuloma: benign pink vascular proliferation at the site of previous trauma or surgery, treated with topical steroids and excision if necessary
Kaposi's Sarcoma: malignant red vascular tumor in patients with AIDS, treated with radiation, excision or AIDS medication (as the immune system improves, the lesions

shrink)

Conjunctival Lymphoma: malignant lymphoma of the conjunctiva, appears pink, treated with radiation

• Degenerative Conditions

These conditions can cause red eye, decreased vision, pain and/or irritation

Pingueculum: a conjunctival bump, may become inflamed, due to UV exposure, treated with lubrication (topical steroids if inflamed)

Pterygium: growth of conjunctiva over the cornea, treated with lubrication or surgical excision if symptomatic or impairing vision (can cause astigmatism or grow across pupil to obstruct vision)

Band Keratopathy: calcium deposition in the cornea due to chronic inflammation or silicone oil, treated with surgical scraping and chelation with EDTA

Dellen: dehydration of part of the cornea, usually associate with a nearby pingueculum or Pterygium, causes red, painful eye, treated with lubrication

Peripheral Ulcerative Keratitis: corneal ulcer due to systemic autoimmune disease (rheumatoid arthritis, lupus, or Wegener's granulomatosis), can cause perforation and blindness, treated with systemic steroids and immunosuppressives, penetrating keratoplasty if perforation occurs

Corneal Dystrophies: collection of disorders that cause deposits in the cornea, can cause decreased vision, treated with lubrication and penetrating keratoplasty if necessary

Fuch's Cornea Dystrophy: a problem with endothelial cells, causes painful corneal edema and slow vision loss, edema is treated with sodium chloride drops and penetrating keratoplasty for advanced cases with vision loss

Keratoconus: bilateral Central corneal thinning and protrusion (cone formation), begins in adolescence, can have painful episodes, treated with hard (rigid gas permeable) contact lenses and penetrating keratoplasty for corneal scarring or contact lens intolerance

Pellucid Marginal Degeneration: bilateral Peripheral (inferior) corneal thinning, causes astigmatism, treated with RGP lenses or penetrating keratoplasty

• Corneal Infectious Conditions

These conditions cause corneal ulceration, redness, pain and loss of vision

Viral Keratitis: due to an immune reaction in the cornea after EKC infection, causes redness, pain, and blurry vision with corneal infiltrates (opacities), treated with topical steroids

Herpes Simplex Keratitis: infection of the cornea due to HSV virus, causes dendrites (branch shaped corneal staining), can also cause iritis, treated with topical trifluridine, oral acyclovir, and topical steroids (when the epithelium is healed)

Varicella-Zoster Keratitis: infection of the cornea with VZV virus, can occur with vesicular rash of the face, treated with acyclovir and later topical steroids, causes neurotrophic cornea (loss of sensation) and dry eye

Bacterial Keratitis: infection with bacteria, causes discharge and corneal ulceration, common in contact lens wearers, treated with culture for bacteria, topical antibiotics, and daily follow-up until the ulcer is healing

Fungal Keratitis: infection with a variety of fungi, also common in contact lens wearers, treated with topical natamycin, fortified amphotericin, and/or systemic antifungals, surgery in severe or non-responsive cases

Acanthamoeba Keratitis: infection with a protozoan, does not respond to typical antibiotics, occurs in contact lens wearer from tap water or swimming in lenses, treated with PHMB, Propamidine, Neomycin, and antifungals

Interstitial Keratitis: opacification of the cornea due to Syphilis or a variety of viruses or bacteria, patients need Syphilis and tuberculosis testing

• Corneal Deposit Conditions

These conditions may be caused by drugs (Amiodarone), diseases (Amyloidosis), or iron (in the tear film), usually no specific treatment is necessary

Fabry's disease: congenital condition that causes corneal deposits (corneal verticillata), associated with cardiovascular, kidney, and CNS disease

• Traumatic Conditions

Thermal Burn: damages the epithelium, causes corneal abrasions, treated with lubrication, topical antibiotics, pressure patching, and cycloplegics (for pain)

Acid Burn: causes corneal opacity and abrasion, protein precipitation limits injury (does not enter the eye), treated with emergency, copious irrigation, topical antibiotics and steroids, frequent lubrication, Vitamin C, and tetracycline

Alkali Burn: worse than acid burns, cause opacity and abrasion, can enter the eye by dissolving the lipid layer of cells, treated as above, remember for both to check the pH after each liter of saline, several liters may be necessary, and irrigation cannot stop until pH is normalized and stable

Traumatic Hyphema: blood in the anterior chamber, due to blunt or penetrating trauma, can cause serious glaucoma, treated with topical cycloplegics, steroids, eye shield, and pressure lowering medications if needed, surgery to wash out if corneal blood staining occurs

Traumatic Iritis: inflammation in the anterior chamber, due to blunt trauma, causes ciliary flush (injection around the cornea) and photophobia, sensitivity to light, treated with topical steroids and cycloplegics

Ghost Cell Glaucoma: high pressure associated with vitreous hemorrhage, treated with pressure lowering drops, wash out surgery or vitrectomy

Lens Subluxation: lens falls (or partially falls) into the vitreous, associate with blunt trauma, causes decreased or fluctuating vision, treated with cataract (lens removal) surgery and vitrectomy

• Lacrimal Conditions

Dacryocystitis: infection of the lacrimal sac (by the nose), due to obstruction of lacrimal drainage, causes pus in the eye and lacrimal sac with red eye, treated with systemic antibiotics, incision and drainage for abscess, and dacryocystorhinostomy

37

(open an new drain from the canaliculi to the nose)

Dacryoadenitis: inflammation of lacrimal gland (by the temple), due to virus or autoimmune disorder (lupus), or TB, causes temporal lid pain and swelling, treated with systemic steroids or antibiotics if infectious

- ## **Scleral Conditions**

Episcleritis: inflammation of the tissue above the sclera, red in color, blanches with phenylephrine, treated with topical steroids or oral NSAIDS (indomethacin)

Scleritis: more serious, associated with rheumatoid arthritis or infection, purple-red in color, exquisitely painful to the touch, can lead to scleral thinning and perforation (ruptured globe), treated with high dose systemic steroids, immunosupressives, and NSAIDS

• SECTION 8: THE LENS, A BRIEF REVIEW

• The Basics

Lens: located behind the iris, becomes more opaque (cataractous) with age, surrounded by a capsule (the capsular bag), accommodates (changes shape) for near vision until 40's and 50's (lose accommodation), and need reading glasses

• Lens Conditions

Coloboma: part of the lens is missing, treated with cataract surgery
Ectopia Lentis: Dislocated lens from birth, treated with cataract surgery
Homocystinuria: systemic disease associated with ectopia lentis, treated with special diet
Traumatic Lens Subluxation: dislocated lens into vitreous associated with trauma, treated with cataract surgery and vitrectomy (because the lens in dislocated into the vitreous)
Pseudoexfoliation: abnormal material deposited on the lens, associated with severe glaucoma, worsens with age, treated with glaucoma drops and surgery

• The Cataract Conditions

Posterior Polar Cataracts: congenital cataracts of the posterior lens, can worsen with age, cause decreased vision, treated with cataract surgery
Nuclear Cataract: slow browning and whitening of the lens as it becomes more and more opaque, treated with cataract surgery when visually significant (impairs the patient's activities of daily living)
Posterior Subcapsular Cataract: disruption and opacification of the posterior lens under the capsular bag, causes glare and decreased vision in bright lights (sunlight), treated with cataract surgery
Other Cataracts: a variety of cataracts due to conditions such as rubella, Wilson's disease, Fabry's and Down Syndrome

• Cataract Surgery

"Extracapsular" Surgery: the capuslar bag is maintained intact so that an intraocular lens (PMMA or Acrylic or Silicone) can be injected into the capsular bag at the end of surgery, both phacoemulsification and extracapsular cataract extraction are types of "Extracapuslar" Surgery
Phacoemulsification: a small corneal incision is made, and ultrasound is used to break up the lens and remove it with a phacoemulsification hand piece, quicker recovery time and shorter visual rehabilitation time (most recent surgery)
Extracapsular Cataract Extraction: a larger incision is made, and the entire lens is expressed out of the capsular bag and out of the eye, then incision is sutured, longer recovery and visual rehabilitation time (more recent surgery)
Intracapsular Cataract Extraction: a large incision, the entire lens and capsular bag are removed together, no lens can be placed in the capsular bag because it is removed,

place an anterior chamber lens or leave the patient aphakic (without a lens) and correct with spectacles

• SECTION 9: RETINA AND VITREOUS, A BRIEF REVIEW

• The Basics

Retina: neural tissue within the globe (eyeball) that senses a light and sends impulses to the optic nerve for the perception of vision

Macula: more central part of the retina, has a larger concentration of cones for the perception of color and sharper vision

Fovea: the most central part of the retina, contains almost entirely cones, responsible for sharpest vision (ability to read letters on a chart)

Peripheral Retina: least central part of the retina, has more rods for dark (dim light vision), and gives peripheral vision

Central Retinal Artery: provides most of the blood supply for the retina, a branch of the carotid artery

Vitreous: a jelly-like material that fills the posterior chamber of the eye, attaches firmly at the optic nerve, vessels, and ora serrata (the termination of the retina in the periphery)

Rods: sense light and dark, for dark-adapted or dim light vision

Cones: sense color, for light-adapted or bright light vision, give sharpest vision (20/20 vision on the chart)

• Retinal Testing

Electroretinogram (ERG): a bright light is flashed under different conditions to stimulate and test rods and cones

Color Vision: colored plates or discs are shown to test cone (color) function, there are three cone types and three types of color deficiency (commonly called "color blindness")

Fluorescein Angiography (FA): dye is injected into a vein, and photos are taken of the retina, if the dye leaks out of the retinal vessels, there is some disease (such as diabetic retinopathy)

Indocyanine Green Angiography (ICG): dye is again injected in the vein and photos are takes, but ICG looks at the choroidal circulation more than FA

Choroidal Circulation: the blood vessels or the choroid (layer beneath the retina), that provide the remaining blood supply to the retina

• Macular Conditions

Age-Related Macular Degeneration (ARMD): condition of age, deposits form below the retina called drusen (yellow colored deposits)

Dry ARMD: only drusen are present, vision is usually good, treated with observation and AREDs vitamins (specially formulated vitamins for macular degeneration)

Wet ARMD: wet because new blood vessels have formed in the area of drusen, these vessels, bleed, leak, and kill the overlying retina causing scarring, treated with laser, Macugen, Lucentis, Avastin (experimental), or photodynamic therapy (PDT)

VEGF: vascular endothelial growth factor, associated with neovascariztion (new vessels) in wet ARMD and with new blood vessels in proliferative diabetic retinopathy (also called neovascularization), promotes vessel growth and leakage of fluid

Wet ARMD Drugs: Macugen, Lucentis, and Avastin are "anti-VEGF" drugs, bind or inhibit VEGF or its receptors, decrease neovascularization (blood vessel growth) and decrease leakage from those vessels in wet ARMD and proliferative diabetic retinopathy

Cystoid Macular Edema: swelling of the macula, associated with cataract surgery (approximately 6 weeks out), diabetes, or uveitis, treated with topical steroid and NSAID drops

Macular Hole: hole in retina right at the fovea, causes severe decreased central vision (peripheral vision is normal), treated with vitrectomy surgery with gas injection, and keeping the patient face down to promote healing

Pathologic Myopia: progressive retinal degeneration in high myopes (> -6.00D), associated with lattice degeneration, retinal holes, and detachment, and macular problems (not ARMD, but a degeneration of the macular all the same), treated with laser for retinal holes or lattice, and retinal detachment surgery if necessary

Photoxic Retinopathy: light kills a portion of the retina, due to looking at the sun, or an extra long ocular surgery (1st year resident doing his first cataract case for 4 or 5 hours), the damage is permanent, treated by prevention (cover the eye when not actually operating)

Drug Toxicity: Tamoxifen (used for breast cancer) causes crystalline deposits in the retina, and chloroquine (used for malaria or lupus) causes macular deposits, scarring, and loss of vision, pts need to check Amsler grid at home and regular eye exam

- ## **Retinal Vascular Conditions**

Diabetic Retinopathy: vascular abnormalities of the retina caused by ischemia (decreased blood flow) to the retina, can be nonproliferative or proliferative, the leading cause of blindness in middle aged Americans

Nonproliferative (Background) Diabetic Retinopathy: characterized by microaneuryms, hemorrhages, retinal edema and exudates (deposits), treated with laser, injection (steroid or anti-VEGF (experimental)), and blood pressure and blood sugar control

Diabetic Exam: every type II (adult onset) diabetic needs an eye exam at the time of diagnosis, every type I (childhood onset) diabetic needs an eye exam within 5 years of diagnosis, both types need an annual exam every year after their initial eye exam

Proliferative Diabetic Retinopathy: new blood vessels form on the retina due to VEGF, new vessels, leak, bleed and cause retinal detachment, treated with laser, vitrectomy, or retinal detachment surgery and control of blood sugar and blood pressure

Focal Laser: laser to the macula for macular edema caused by nonproliferative diabetic retinopathy, the microaneuryms leak, and the laser seals the microaneuryms and stops the leakage

Panretinal Photocoagulation (PRP): laser to the peripheral retinal for proliferative diabetic retinopathy, the ischemic retina in the periphery is creating VEGF and laser kills the peripheral retina, stopping VEGF production, and stopping vascular proliferation (the neovascularization regresses)

Sickle Cell Retinopathy: sickle cell disease causes retinal ischemia and neovascularization much like diabetes, can be observed or treated with laser if necessary

Hypertensive Retinopathy: high blood pressure causes leakage and bleeding from the retinal vessels, cotton-wool spots and hemorrhage seen on exam, treated with blood pressure control

Central Retinal Vein Occlusion (CRVO): associated with hypertension and vascular disorders, vein becomes occluded, leads to increased pressure, bleeding and leakage in the retina

Central Retinal Artery Occlusion (CRAO): associated with atherosclerosis and cholesterol, the artery becomes occluded, blood flow stops, and the retina dies due to ischemia, treated with glaucoma drops to lower pressure and carbogen (carbon dioxide/oxygen combination)

Retinopathy of Prematurity: neovasc016ztion of the retina in low birth weight, premature infants, causes retinal detachment in infants, treated with exams every 2 weeks and PRP or retinal detachment surgery if necessary

• Retinal Inflammatory/Infectious Conditions

Toxoplasmosis: infection of the retina caused by Toxplasma gondii, transmitted by cat feces, treated with systemic trimethoprim/sulfamethoxazole, sulfadiazine, pyrimethamine, and systemic steroids (prednisone)

Sarcoidosis: systemic disease with inflammation in the lungs, skin, joints, and eyes, usually in African Americans less than 40, causes iritis and dacryoadenitis, treated with systemic and topical steroids, and immunosuppressives

Other Retinochoroidopathies: a variety of diseases with inflammation of the retina, fluorescein angiogram is useful in making the diagnosis

Vogt-Koyanagi-Harada Syndrome: bilateral uveitis associated with meningitis, usually in Asian, Hispanics, or Native Americans, autoimmune, associated with madarosis (white lashes), vitiligo (white skin patches), treated with systemic steroids and immunosuppressives

Acquired Immunodeficiency Syndrome (AIDS): HIV infection causing decreased immunity, pts get HIV retinopathy (hemorrhage and cotton-wool spots), and infectious retinopathy (Cytomegalovirus (CMV), Toxoplasmosis, and Herpes), treated with HIV medications anti-infectives for specific diseases, pts with AIDS and low CD4 (certain white blood cell) cell count (50 or below) should have a dilated fundus exam immediately, and every 3 months thereafter, and come in immediately for any floaters or change in vision

Syphilis: infection with Treponema pallidum causes keratitis, iritis, and retinitis, treated with IV penicillin for 10 days (called the neurosyphilis regimen) because the eye is like an extension of the brain, should have lumbar puncture to test for neurosyphilis as well

Lyme Disease: infection with Borrelia burgdorferi, causes uveitis, treated with doxycycline

Cat-Scratch Disease: infection with Bartonella henselae, causes inflammation of the nerve and retina (hence neuro-retinitis), treated with erythromycin or doxycycline

Cytomegalovirus (CMV) Retinitis: viral retinitis in AIDS patients with CD4 count less than 50-100, causes blindness and retinal detachment, treated with IV ganciclovir or with surgery to place a ganciclovir implant in the vitreous that releases the medication over several years

• Hereditary Retinal Conditions

Best's Disease (Vitelliform Dystrophy): inherited macular disease where pts lose central vision slowly, the EOG (Electro-Oculogram) is abnormal (really the only time you may see the EOG used), no good treatment

43

Stargardt's Disease: inherited macular disease where pts lose central vision, diagnosed with fluorescein angiogram, no good treatment

X-Linked Juvenile Retinoschisis: inherited retinal disease that can lead to retinal detachment, but usually false retinal detachment called schisis (the layers of the retinal are separated from themselves), treated with retinal detachment surgery if it occurs

Retinitis Pigmentosa: inherited condition where the photoreceptors (the rods mainly) in the retina die, and black, retinal scars occur (called bone spicules), ERG (electroretinogram) is diagnostic, showing that the rods are damaged, acetazolamide can be used for macular edema associated with RP

Albinism: inherited loss of pigment in the eyes or the eyes and whole body, pts have poor vision and nystagmus from infancy, treated with UV blocking sunglasses, sun avoidance, low-vision training, and genetic counseling if considering children

The Metabolic Diseases: a group of diseases from an enzyme deficiency, can cause a "cherry-red spot" appearance to the macula, treated by treating the underlying genetic disease, if possible

- ## Peripheral Retina and Vitreous Conditions

Posterior Vitreous Detachment: with age the vitreous in the eye liquifies, it is surrounded in a sac called the hyaloid, the hyaloid sac can then pull away from the retina causing flashing lights, and the condensed portions of the hyaloid cause floater (black moving spots in the vision), 85% of pts have no problems, and the floaters decrease in size with time, 15% may have retinal holes or tears, treated with laser retinopexy (laser around the hole to prevent detachment) if holes are present, pts should have exam in 4 to 6 weeks, and retinal detachment precautions (return immediately for new floaters or loss of vision)

Lattice Degeneration: a weakening in the peripheral retina, associated with high myopia, can lead to retinal holes and detachment, treated with observation, laser around holes (retinopexy), or retinal detachment surgery if it occurs (less than 0.5%)

Retinal Holes and Tears: seen on indirect ophthalmoscopy, treated with laser retinopexy, can be treated with cryotherapy in the OR

- ## Retinal Detachment Conditions

Rhegmatogenous Retinal Detachment: a hole occurs in the retina, liquified vitreous flows into the hole and detaches the retinal from the underlying choroid, retinal cells die if detached from the choroidal blood supply too long, treated with retinal detachment surgery (scleral buckle surgery is good for this type of detachment)

Traction Retinal Detachment: a neovascular membrane (say from proliferative diabetes or retinopathy of prematurity) pulls on the retina as it grows, pulling the retina off the choroid, treated with retinal detachment surgery (pars plana vitrectomy is good for this type of detachment because the surgeon must get inside the eye and remove the neovascular membrane's attachments to the retina or vitreous, to relieve the traction)

Exudative Retinal Detachment: occurs with inflammation in the eye that causes fluid (called exudate) to form between the retina and choroid, treated by treating the underlying inflammation (usually not surgery)

Proliferative Vitreoretinopathy (PVR): proliferation of cells and fibrosis after retinal surgery, causes the retina to re-detach (number one cause of retinal detachment surgery failure), difficult to fix, treated with more retinal detachment surgery (pars plana vitrectomy)

Intraocular Gasses and Oils: injected into the eye at the end of retinal detachment surgery to keep the retina pushed against the choroid (keep the retina attached), Air lasts a few days, C3F8 gas lasts 8 weeks, and Silicone oil is permanent, the gases dissolve on their own, silicone oil must be removed with another surgery, pts with air in their eye CANNOT fly or have inhalational anesthetics

- ## Trauma to the Retina and Vitreous

Commotio Retinae: whitening of the retina in an area of blunt-force trauma, causes temporary decreased vision, resolves without treatment

Traumatic Macular Hole: retinal hole in the center of the fovea due to blunt-force trauma, causes severe decreased vision, treated with vitrectomy and intraocular gas injection

Traumatic Retinal Tears and Holes: holes from trauma, treated with laser retinopexy

Suprachoroidal Expulsive Hemorrhage: may occur from penetrating trauma, or during almost any kind of ocular surgery (the MOST dreaded complication of the ocular surgeon), a hemorrhage of the posterior ciliary artery causes explosion of all the ocular contents out of the eye (through the surgical or traumatic wound), causes sudden, untreatable, irreversible blindness.

Intraocular Foreign Body: something stuck inside the eye that shouldn't be there, a surgical emergency, from high speed injuries (circular saw or hammering), foreign body must be surgically removed as soon as possible, and the ruptured globe (perforation injury) must be closed, these pts may need topical and intraocular antibiotics

Endophthalmitis: infection within the eye after penetrating trauma or surgery (the second most dreaded complication of the ocular surgeon), from implanted bacteria or fungi, causes redness, pain, decreased vision, treated with topical and intraocular antibiotics, closure of a ruptured globe if necessary, vitrectomy if severe, can cause permanent blindness and loss of the eye

Sympathetic Ophthalmia: inflammation in the second eye (normal eye) after trauma or disease to the first eye (the sick eye), treated with topical and systemic steroids and enucleation of the first (sick) eye

- ## Retinal Tumor Conditions

Choroidal Melanoma: malignant tumor from the choroid or ciliary body, can be asymptomatic or cause decreased vision, diagnosed with exam, ultrasound and fluorescein angiogram, treated with radioactive plaque insertion or enucleation

Choroidal Nevus: benign tumor of the choroid, remains flat, does not grow like a melanoma

Intraocular Lymphoma: malignant lymphoma of the eye, causes inflammation (uveitis) and can cause a mass, and is treated with chemotherapy and steroids

Tuberculoma: an inflammatory mass in the retina or choroid that is an inflammatory nodule from infection with tuberculosis

• SECTION 10: UVEITIS, A BRIEF REVIEW

• The Basics

Uveitis: inflammation within the eye, can be divided into anterior, intermediate, posterior, and panuveitis
Anterior Uveitis: inflammation in the iris and anterior chamber of the eye
Intermediate Uveitis: inflammation around the pars plana (area of the eye from behind the lens to the anterior beginnings of the retina), causes cells in the vitreous
Posterior Uveitis: inflammation in the vitreous, retina, and choroid
Panuveitis: inflammation of the entire eye, a combination of anterior, intermediate, and posterior uveitis
Scleritis: inflammation of the white scleral coat of the eye, may be isolated or in combination with other uveitis (redness of scleritis does not blanch with phenylephrine)
Assessment of Uveitis: inflammation causes red eye and white blood cell collections in the anterior chamber, vitreous and retina or choroid, white cells can be seen at the slit lamp at very high magnification to grade uveitis, and flourescein angiogram may also be helpful for posterior uveitis
Uveitis Treatment: usually includes the use of steroids (drops or systemic) to reduce inflammation and cycloplegics to reduce photophobia (pain with looking at light)

• Anterior Uveitis Conditions

HLA-B27 Anterior Uveitis: autoimmune disease where uveitis is associated with inflammation in other parts of the body (arthritis, low back pain, Crohn's disease) treated with topical steroid drops
Posner-Schlossman Syndrome: anterior uveitis associated with high eye pressure, treated with both topical steroid and anti-glaucoma drops
Phacoantigenic Uveitis: anterior uveitis associated with trauma to the crystalline lens, treated with steroid drops and removal of the lens
Viral Anterior Uveitis: from infection with Herpes zoster or Herpes simplex virus, can cause high pressure, treated with topical steroids and systemic antivirals (acyclovir)
Juvenile Rheumatoid Arthritis: anterior uveitis that begins in childhood, associated with arthritis and joint pain, treated with topical and systemic steroids, and systemic immunosuppression
Traumatic Iritis: anterior uveitis after blunt injury to the eye, treated with a short course of topical steroid drops and cycloplegics
Fuch's Heterochromic Iridocyclitis: chronic anterior uveitis in one eye that eventually causes that eye to look lighter in color (more blue than the normal eye)

• Intermediate Uveitis Conditions

Intermediate Uveitis in Brief: causes floaters in the vitreous (vitreous cells), the eye may appear non-inflamed with white sclera, due to autoimmune or infectious conditions
Sarcoidosis: causes anterior, intermediate or posterior uveitis, usually in African Americans, associated with skin nodules and lung disease, pts have high ACE level, treated with topical or systemic steroids, and immunosuppressives
Tuberculosis: from Mycobacterium tuberculosis, causes all types of uveitis, pts have

positive PPD, treated with systemic, anti-TB meds (Isoniazid, Rifampin, Ethambutol and others), steroids are contraindicated in pts with active TB (can make the lung infection even worse)

Syphilis: from Treponema pallidum, causes all types of uveitis, can be congenital (passed from birth mother to fetus) or acquired, treated with IV penicillin, congenital syphilis can be associated with abnormal teeth, nose and other development

Pars Planitis: chronic intermediate uveitis of unknown cause, usually bilateral, can be treated with steroids, immunosuppressives or pars plana vitrectomy

Lyme Disease: causes many types of uveitis, treated with doxycycline or ceftriaxone, pts have a characteristic "bull's eye" rash after being bitten by the tick infected with bacterium (Borrelia burgdoferi)

- ## Endophthalmitis Conditions

Acute Postoperative Endophthalmitis: usually due to Staph. aureus infection, severe inflammation within 1-2 days of surgery, painful, red eye with decreased vision, due to infection, treated with intraocular antibiotic injections, pars plana vitrectomy, and fortified topical antibiotics

Chronic Postoperative Endophthalmitis: usually due to Staph. epidermidis or Proprionibacterium acnes, causes chronic, low-grade inflammation, treated with intravitreal antibiotics, surgery to remove any retained material (left over from the cataract procedure), or vitrectomy

Bleb-Associated Endophthalmitis: usually due to Streptococcus or Haemophilus, begins with infection of a glaucoma filtering bleb (the bleb becomes white and full of pus and the eye becomes very red), spreads to the anterior chamber, then the posterior chamber, and then the entire eye, treated with fortified topical antibiotics for infection of the bleb only (called blebitis), or intraocular antibiotics for more serious infection

Posttraumatic Endophthalmitis: infection after penetrating trauma deposited organisms inside the eye, treated with intraocular antibiotics and vitrectomy

Endogenous Endophthalmitis: infection of the eye spread from the patients own blood (they have the infection somewhere else in their body), occurs in pts with indwelling venous catheters or severe systemic infection, treated with systemic antibiotics, catheter removal, intraocular antibiotics and vitrectomy

- ## Posterior Uveitis Conditions

White Dot Syndromes: group of disorders that cause retinal and choroidal inflammation, diagnosed with indirect exam and fluorescein angiogram, sometimes treated with steroids

Ocular Toxoplasmosis: infection with Toxoplasma gondii, spread from cat feces, causes retinitis and vitritis, (sometimes the only sign of infection is an old retinal scar that is inactive), treated with systemic trimethoprim/sulfamethoxazole, sulfadiazine, or clindamycin, oral steroids are added after antibiotics have been started

Ocular Toxocariasis: infection with Toxocara canis, spread from dog feces, no good treatment

Presumed Ocular Histoplasmosis Syndrome (POHS): infection with Histoplasma capsulatum (fungus), can cause choroidal neovascularization (CNV), the CNV is treated with laser or anti-VEGF injections

Behcet's Disease: autoimmune, causes all types of uveitis, associated with oral and genital ulcers, usually pts from Asia or Turkey, treated with systemic steroids and immunosuppressives

Sympathetic Ophthalmia: inflammation in the second eye (normal eye) after trauma or disease to the first eye (the sick eye), treated with topical and systemic steroids and enucleation of the first (sick) eye

Vogt-Koyanagi-Harada Syndrome: bilateral uveitis associated with meningitis, usually in Asian, Hispanics, or Native Americans, autoimmune, associated with madarosis (white lashes), vitiligo (white skin patches), treated with systemic steroids and immunosuppressives

Intraocular Lymphoma: causes inflammatory cells in the vitreous and retina, associated with systemic lymphoma, treated with radiation and chemotherapy

- ## AIDS Associated Conditions

AIDS in General: viral infection with HIV, causes decreased T-Cell immunity, monitor the patients immunity with CD-4 cell count (a type of T-cell), predisposes the patient to a variety of infectious and non-infectious eye diseases, pts with AIDS need routine dilated fundus exam at least yearly (and up to every 3 months if CD-4 count is less than 50), and should be instructed to come in immediately for any change in vision or floaters

HIV Retinopathy: retinal changes with cotton-wool spots and hemorrhages, treated with systemic anti-AIDS medication combinations (called HAART- Highly Active Anti-Retroviral Therapy)

CMV Retinitis: from infection with Cytomegalovirus, causes progressive retinal hemorrhage and death, can lead to total blindness and retinal detachment if untreated, treated with anti-virals (ganciclovir or valganciclovir), surgery to implant a ganciclovir implant to slowly release the drug, or retinal detachment surgery

Kaposi's Sarcoma: infection with Herpes virus 8, causes red-colored tumors on the conjunctiva and skin, treated with HAART, surgical excision, chemotherapy, and/or radiation

Herpes Zoster Ophthalmicus: reactivation of Herpes Zoster (aka Varicella Zoster) virus (same virus that causes chicken pox), occurs in the V1 dermatomal distribution that includes the tip of the nose, eye, and forehead, is always unilateral (affecting one side of the face only), causes many forms of uveitis, treated with acylovir

Other Conditions: pts with AIDS are also more likely to develop toxoplasmosis, fungal endophthalmitis, Tuberculosis, and a variety of infectious retinal and skin conditions

Rifabutin Anterior Uveitis: inflammation of the anterior chamber from taking Rifabutin (a drug for Tuberculosis treatment), treated by stopping the drug

Cidofovir Anterior Uveitis: inflammation from Cidofovir injection (drug used to treat CMV retinitis), cause inflammation and irreversible hypotony, treated by discontinuing the drug, damage may be permanent

• SECTION 11: GLAUCOMA, A BRIEF REVIEW

• The Basics

General: the eye is a pressurized sphere (like a balloon), without some pressure the eye would collapse (like a deflated balloon)

Aqueous Humor: the fluid that is constantly produced by the ciliary body in the posterior chamber, flows through the pupil to the anterior chamber, and is absorbed into the blood stream through the trabecular meshwork

Ciliary Body: located behind the iris, produced aqueous humor

Trabecular Meshwork: a filter located at the anterior chamber angle (area where the cornea slopes down to meet the anterior iris), removes aqueous from the anterior chamber into the blood stream

Intraocular pressure (IOP): the pressure within the eye that is determined by the production and removal of aqueous humor, measured with tonometry

Tonometry: measuring pressure through contact with the eye (Goldmann or Tono-Pen contact tonometry), or without contact (air-puff non-contact tonometry)

Pachymetry: ultrasound measurement of the thickness of the cornea in its very center, important as a risk factor for glaucoma (pts with a thin cornea have higher risk of glaucoma)

Gonioscopy: examination of the anterior chamber angle with a prism to see if if is open or narrow/closed (pts with a narrow/closed angle are at risk for narrow angle glaucoma)

Perimetry: examination of the patient's peripheral vision (glaucoma can affect the peripheral vision before the pt's central vision is lost)

Goldmann Perimetry: a technician moves a light to measure the peripheral vision, this is manual (a human does the test), dynamic (the light is moving) perimetry

Automated Perimetry: the computer flashes lights to measure peripheral vision, this is automated (a computer does the test), static (the light does not move) perimetry, like a Humphrey visual field

Optic Disc Cupping: the optic nerve as viewed inside the eye looks like a donut, the center of the donut is normally about one third of the donut (a healthy disc without cupping), in more advanced glaucoma, the nerve fibers die and the cup becomes larger than one-third of the disc (the donut has a bigger and bigger hole), when the cup is more than a third of the disc, this is called optic disc cupping

Disc Imaging: includes photos of the optic disc to look for changes in cupping over time, OCT (infrared analysis of thickness) of the nerve fiber layer to look for nerve fiber death, and HRT (scanning laser analysis) that can measure disc cupping

Glaucoma Diagnosis: diagnosing a pt with "glaucoma" is difficult, it includes a combination of factors and tests that, when all analyzed together, make the diagnosis of glaucoma, these factors include intraocular pressure, visual field loss on perimetry, disc cupping on exam or HRT, and nerve fiber layer loss on exam or OCT

Scotoma: loss of vision in part of the pts visual field as measured by perimetry

Glaucoma Risk Factors: Age, high intraocular pressure, thin cornea on pachymetry, family history, African American race, and others, these should be assessed in every potential glaucoma patient

Glaucoma Treatment: of the above risk factors, intraocular pressure is the only one we can control, so almost all glaucoma treatment focuses of lowering the intraocular pressure with topical anti-glaucoma drops, glaucoma laser procedures, or glaucoma surgical procedures

End Stage Glaucoma: as visual field loss progresses, the patient can become completely blind from glaucoma (having no light perception)

"The Glaucomas": there are a variety of different types of glaucoma with differing etiologies, usually the patients have some degree of nerve damage (cupping) that can occur with or without high pressure

Optic Neuropathy: damage to the optic nerve that can occur from a variety of reasons, one reason is glaucoma (as the nerve fibers die) causing cupping

Following the Glaucoma Patient: IOP should be checked each visit (at least every 3 to 6 months), photos should be taken and compared annually, perimetry should be performed and compared every 6 to 12 months, any of these things more frequently if the disease is progressing

Glaucoma Progression: worsening of disc cupping and visual field loss on perimetry

Goal IOP: a level of intraocular pressure set by the physician for the patient undergoing glaucoma treatment with drops or after surgery, if the patient continues to have glaucoma progression despite having a lower IOP at their goal IOP, the goal may be changed to a lower goal

Open-Angle Glaucoma: glaucoma in pt that does NOT have narrow angles on gonioscopy (includes many different subtypes)

Narrow-Angle Glaucoma: glaucoma in a pt that DOES have narrow angles (or closed angles) on gonioscopy

Narrow-Angle: anatomic description of the angle formed where the iris meets the peripheral cornea, can be caused by persons anatomy, a big lens, or because of high pressure of the aqueous in the posterior chamber that pushes the iris forward

Glaucoma Suspect: the pt with some signs of glaucoma, but no definite signs of optic neuropathy, these pts are followed closely with testing and IOP measurements

Ocular Hypertension: high eye pressure (over 21mm Hg) without any signs of glaucomatous optic neuropathy, pts with high pressure are called ocular hypertensives, and are either followed closely with testing or treated with IOP lowering drops to normalize the pressure

Hypotony: low eye pressure that can lead to macular folds and permanent vision loss, can occur after glaucoma surgery (too much aqueous leaves the eye)

- **Treatment of Open-Angle Glaucoma**

Topical Anti-Glaucoma Medications: (beta blockers, carbonic anhydrase inhibitors, prostaglandin analogues, adrenergic agonists, and cholinergic agents) drops that lower the IOP by decreasing the production of aqueous or increasing the outflow of aqueous into the bloodstream

Systemic Anti-Glaucoma Medications: (carbonic anhydrase inhibitors) pills that lower the IOP by decreasing the production of aqueous

Laser Trabeculoplasty (ALT/SLT): laser applied to the trabecular meshwork to increase outflow of aqueous through the trabecular meshwork into the bloodstream

Trabeculectomy: surgery to open a new escape route for aqueous from the anterior chamber to the subconjunctival space

Tube Shunt: surgery to place a tube in the anterior chamber that leads to the subconjunctival space, the tube is the new escape route for aqueous

Goniotomy/Trabeculotomy: surgery to the trabecular meshwork to open it for better aqueous flow (used in infants with congenital glaucoma)

- ## Treatment of Narrow-Angle Glaucoma

Laser Iridotomy: laser used to drill hole in the iris to allow aqueous to flow from the posterior to the anterior chamber, this should equalize the pressure between the two chambers and allow the iris to fall backward against the lens, changing the narrow angle to a normal/open angle configuration, this is used to prevent an acute angle-closure glaucoma attack

Laser Iridoplasty (PICP): laser applied to the peripheral iris 360 degrees to pull the iris away from the angle and relieve the narrow angle, used in patients with persistent narrow angles because of anatomy (called plateau iris)

- ## Open-Angle Glaucoma Conditions

Primary Open-Angle Glaucoma (POAG): optic neuropathy with cupping due to glaucoma with no underlying cause, associated with older age and higher eye pressure, treated by lowering the IOP with drops, laser, or surgery

Normal Tension Glaucoma: progressive optic neuropathy with cupping in a patient with normal eye pressure, treated as POAG

Pigment Dispersion Syndrome: open angle glaucoma with lots of iris pigment in the anterior chamber and stopping up the trabecular meshwork, treated as POAG, but may also do laser iridotomy

Pseudoexfoliation Syndrome: open angle glaucoma with abnormal material deposits on the lens, treated as POAG, pts have abnormal zonules and difficult cataract surgery

- ## Angle-Closure Glaucoma Conditions

Primary Angle-Closure Glaucoma: narrow angle glaucoma caused by the iris blocking the trabecular meshwork, no aqueous can exit the eye and pressure builds, can be acute, intermittent or chronic

Acute Angle-Closure Glaucoma: the pt has a sudden attack with severe pain, vomiting, decreased vision, and a red eye, caused by increased aqueous pressure in the posterior chamber pushing the iris forward and closing the angle leading to a sudden increase in IOP, treated with IOP lowering drops and pills, then a laser iridotomy as soon as possible

Intermittent Angle-Closure Glaucoma: pts have occasional symptoms when the angle closes temporarily due to lighting changes, pts need laser iridotomy to prevent an acute angle-closure attack

Chronic Angle-Closure Glaucoma: pts have chronic elevated pressure due to chronic angle closure, may not have an acute angle-closure attack, treated with laser iridotomy and IOP lowering drops

Secondary Angle-Closure Glaucoma: narrow angle glaucoma caused by another condition such as a large cataractous lens or neovascularization of the iris, treated with glaucoma drops, surgery, and laser

Neovascular Angle-Closure Glaucoma: pts with proliferative diabetic retinopathy can have neovascularization of the iris as well as the retina, blood vessels grow over the trabecular meshwork and close the angle, this is treated acutely with pressure lowering drops and pills, but PRP laser to the retina is necessary to stop the neovascularization (the blood vessels will regress after laser treatment)

- ## Secondary Glaucoma Conditions

Secondary Glaucoma: glaucoma caused by an underlying condition with high pressure

Uveitic Glaucoma: glaucoma associated with underlying uveitis (from many causes), treated by treating the uveitis with steroids or immunosuppression and treating the glaucoma simultaneously with drops or surgery

Posner-Schlossman Syndrome: glaucoma associated with uveitis of unknown cause, treated by treating the inflammation and glaucoma simultaneously with drops or surgery

Fuch's Heterochromic Iridocyclitis: unilateral glaucoma and uveitis that changes lightens the color or the effected eye (turns more blue)

Phacolytic Glaucoma: glaucoma due to an overmature cataract that is disintegrating within the eye and releasing lens proteins that block the trabecular meshwork, treated with cataract extraction

Lens Particle Glaucoma: glaucoma due to left over lens fragments after cataract surgery that clog the trabecular meshwork, treated with anti-glaucoma drops or surgery, or surgery to remove the lens particles

Phacoanaphylactic Glaucoma: uveitis and glaucoma due to lens trauma, an autoimmune reaction to the lens, treated with cataract extraction and anti-glaucoma drops or surgery, and steroid drops

Hyphema-Associated Glaucoma: red blood cells in the anterior chamber clog the trabecular meshwork, treated with IOP lowering drops, or surgery to wash out the blood from the anterior chamber

Ghost Cell Glaucoma: old red blood cells that have turned white with time (called ghost cells), clog the trabecular meshwork, treat with IOP lowering drops or surgery to wash our the anterior chamber or vitrectomy to clean out the old red blood cells in the vitreous (the original location of the hemorrhage)

Angle Recession Glaucoma: blunt trauma to the eye rips part of the iris root (attachment) at the anterior chamber angle, the damaged angle can lead to glaucoma years later, treated as POAG

Steroid-Induced Glaucoma: high pressure (over 21) due to topical or systemic steroid use, a common response in pts with POAG, treated by stopping steroids and using IOP lowering drops or surgery

Malignant Glaucoma: glaucoma after IOP lowering surgery, due to aqueous flowing into the vitreous and causing angle closure, treated with laser iridotomy, IOP lowering drops, and vitrectomy

Iridocorneal Endothelial (ICE) Syndrome Glaucoma: glaucoma due to abnormal endothelial cells that grow over the trabecular meshwork, treated with IOP drops or surgery

Congenital Glaucoma: glaucoma in infants due to congenital abnormality of the angle, treated with IOP drops or angle surgery (Goniotomy or Trabeculotomy), these children present with photophobia, pain, and a whitened cornea (corneal edema)

- ## SECTION 12: NEUROOPHTHALMOLOGY, A BRIEF REVIEW

  - ## The Basics

Neuroophthalmology: includes vision problems from the optic nerves to the brain, and motility problems related to the cranial nerves that control the eye muscles
Neuroophthalmologic Exam: includes visual acuity, color vision testing, pupil testing, motility testing, and dilated fundus exam to look at the nerves; as well as a complete neurologic exam
Swinging Flashlight Test: a light is shone into each eye back and forth, normally each pupil constricts equally when the light goes to each eye
Afferent Pupillary Defect: on swinging flashlight test, one pupil constricts (normal one), the other pupil dilates (the abnormal one) in the eye with optic neuropathy
Optic Neuropathy: damage to one optic nerve due to a variety of causes, leads to afferent pupillary defect, color vision loss, and visual field loss
Functional Vision Loss: code word for malingering (pretending visual loss for some kind of secondary gain, such as disability), these pts often end up with the neuroophthalmologist for testing
CT Scan (Computed Tomography Scan): type of X-ray used to look for brain or optic nerve tumors that can affect vision or the optic nerve (pituitary tumor or optic nerve sheath meningioma)
MRI Scan: type of X-ray used to look for inflammation in the optic nerve (optic neuritis) or the brain (as seen in multiple sclerosis)
Optic Neuritis: inflammation of the optic nerve due to a variety of causes that leads to optic nerve damage (optic neuropathy)
Visual Field Defects: tested for with perimetry, can have a specific pattern that indicates a specific location of damage in the brain or optic nerve or optic tract
Occipital Lobe: portion of the brain responsible for fine vision
Protective Lenses: spectacles made of a special plastic called polycarbonate, that is shatter-proof, to be worn full-time by patients with only one good eye (the other eye has severe disease), for many patients with amblyopia or permanent optic nerve damage, the only and most important intervention you can offer is protective lenses to help protect against damage to the better eye

  - ## Optic Nerve Head Conditions

Optic Nerve Coloboma: congenital absence of part of the optic nerve, associated with decreased vision in one eye, no treatment
Optic Pit: congenital defect in a small part of the optic nerve, may be asymptomatic or have a small scotoma, no treatment
Optic Nerve Hypoplasia: congenitally small, hypoplastic optic nerve, associated with low vision and nystagmus, small nerve seen on exam, MRI or CT scan, no treatment
Septo-Optic Dysplasia (De Morsier Syndrome): optic nerve hypoplasia with absence of the septum pellucidum (midline brain structure), associated with pituitary abnormalities and hormone problems, treated with hormone replacement
Tilted Optic Disc: oval shape to the optic nerve head seen in high myopes, no

treatment

Papilledema: swelling of the optic nerve due to high intracranial pressure (pressure in the brain ventricular system filled with CSF), caused by conditions that increase intracranial pressure (tumor, hydrocephalus, or pseudotumor), patients need CT or MRI of the brain looking for tumor, lumbar puncture to measure CSF pressure, surgery for tumors or acetazolamide for pseudotumor

Pseudotumor Cerebri: high intracranial pressure without tumor or underlying cause, usually in obese females, treated with oral acetazolamide and weight loss, or surgery (optic nerve sheath fenestration to cut holes in the lining that covers the optic nerve called the optic nerve sheath, to release the pressure)

Optic Nerve Head Drusen: deposits in the optic nerve head, looks like papilledema, usually asymptomatic, pts need perimetry, no treatment

Melanocytoma: congenital pigment of the optic nerve head, usually asymptomatic, follow with photographs for growth, no treatment

Myelinated Retinal Nerve Fiber Layer: whitish appearing nerve fibers in the retina (feathery in appearance on exam), usually asymptomatic, no treatment

• Optic Nerve and Chiasm Conditions

Ischemic Optic Neuropathy: loss of blood supply to the optic nerve that kills the optic nerve, can be due to giant cell arteritis or atherosclerosis

Giant Cell Arteritis: inflammatory disorder of vessels to the head (temporal artery), and the eye, diagnosed with high ESR test and biopsy of the temporal artery, treated with systemic steroids

Optic Neuritis: inflammation of the optic nerve, causes swelling (edema) of the nerve, seen on clinical exam or with MRI, occur alone or in association with multiple sclerosis, usually improves without treatment, can use IV then po steroids

Nutritional or Toxic Optic Neuropathies: occur with vitamin deficiency (B12 or Folate) or with toxins (many meds, especially ethambutol for TB or vincristine for cancer), treated with vitamins or toxic drug cessation

Leber's Hereditary Optic Neuropathy: inherited optic neuropathy, usually in men in their twenties, they have rapid vision loss, no good treatment

Inflammatory Optic Neuropathy: due to inflammatory disease or infection such as sarcoidosis, tuberculosis, or syphilis, treated by treating the underlying disease

• Optic Nerve Tumor Conditions

Optic Nerve Glioma: tumor of the optic nerve, usually in children, diagnosed on MRI or CT, treated with observation (serial MRI) and if growing, excision of tumor with enucleation, possible radiation or chemotherapy

Optic Nerve Sheath Meningioma: tumor or the optic nerve sheath (extension of the dura mater from the brain that surrounds the optic nerve), diagnosed on MRI or CT, treated with excision and enucleation

Traumatic Optic Neuropathy: damage to the nerve due to trauma, treated with steroids

• Extraocular Muscle Conditions

Third Nerve Palsy: damage to cranial nerve 3 (weakens all muscles except superior oblique and lateral rectus), eye is turned "down and out" on exam, causes diplopia (double vision), if the pupil is involved (a dilated, nonreactive pupil) suspect brain aneurysm, otherwise can be due to a small stroke to the nerve (atherosclerosis usually in pts with diabetes and hypertension), MRI and MRA imaging is recommended to rule out aneurysm, motility improves over six months, may need strabismus surgery
Brown Syndrome: the eye cannot elevate or adduct, due to a superior oblique tendon problem (not a nerve), usually congenital, treated with strabismus surgery
Superior Oblique Myokymia: the superior oblique jumps repetitively causing the pt to complain or the world rotating or shaking (called oscillopsia), treated with carbamazepine or strabismus surgery
Sixth Nerve Palsy: damage to cranial nerve 6 (weakens the lateral rectus muscle), eye is turned inward toward the nose, causes diplopia (double vision), usually due to atherosclerosis, pts need MRI to rule out tumor, improves over six months, treated with strabismus surgery if necessary
Fourth Nerve Palsy: damage to cranial nerve 4 (weakens the superior oblique muscle), eye is turned upward, causes vertical diplopia (double vision up and down), can be congenital or due to atherosclerosis, MRI needed in adults, improves over six months in adults, treated with strabismus surgery if necessary
Multiple Cranial Nerve Palsies: sign of brainstem, brain or orbital disease, these pts need an MRI
Cavernous Sinus Syndrome: cranial nerves III, IV, V1, VI, and VII all pass through or close to the cavernous sinus (a vein in the brain), and can be affected together by tumor in this area or carotid-cavernous fistula (arterio-venous shunt)
Chronic Progressive Ophthalmoplegia (CPEO): muscle disease with progressive paralysis of the extraocular muscles, can be associated with cardiac problems (called Kearns-Sayer Syndrome) and pts need a cardiac evaluation
Marcus-Gunn Jaw Winking Syndrome: pts have ptosis of one eye, the eye opens more when the jaw is opened, usually congenital, no treatment
Duane Syndrome: inability to abduct or adduct the eye, usually congenital, can be treated with strabismus surgery
Aberrant Regeneration: after third nerve palsy, the nerve re-connects inappropriately causing lid retraction or abnormal movements
Internuclear Ophthalmoplegia: one eye will not move past midline, due to lesion in the medial longitudinal fasiculus in the brainstem
Dorsal Midbrain Syndrome: pts have nystagmus, upgaze paresis, and light-near dissociation of the pupils (constrict at near but not to light), usually due to tumor near dorsal midbrain (pinealoma)
Seventh Nerve Palsy: damage to cranial nerve 7 (facial nerve), causes facial weakness and droop, caused by brain lesion usually, need MRI scan
Nystagmus: rhythmic eye movement beating to one side or the other, can be due to congenital poor vision, loss of vision, or brain lesions, most patients need an MRI scan, there are many types of nystagmus associated with different diseases and lesions

- **Pupil Response Conditions**

Afferent Pupil Defect: called Marcus-Gunn Pupil, swinging flashlight test shows dilation in one eye (abnormal eye), and constriction in the other (normal eye), both pupils will constrict at near (part of accommodation), usually a sign of damage to the optic nerve

Anisocoria: one pupil is larger in size than the other, can be due to Horner's syndrome, pharmacologic dilation (especially in healthcare or eyecare workers), or physiologic

Horner's Syndrome: pts have ptosis (lower lid on one side), miosis (a smaller pupil of one side), and anhydrosis (do not sweat on one side of the face), congenital or due to lung tumor that damages the sympathetic nerves as they travel from the spine to the head through the lungs, pts may need MRI of brain and spinal cord

Adie's Tonic Pupil: chronically constricted pupil, may be due to infection (such as syphilis), pts need syphilis testing and treatment if positive

Third Nerve Palsy: classified as "pupil sparing," a motility problem without pupil involvement, or "pupil involving," with a dilated, unreactive pupil, due to brainstem herniation or aneurysm

Pharmacologic Dilation: isolated dilated pupil with no other findings, usually in healthcare or eyecare personnel

- ## Systemic Conditions

Multiple Sclerosis: demyelinating disease of the CNS, can affect cranial nerves and cause optic neuritis, attacks are treated with steroids, interferon and copaxone are also used

Thyroid Eye Disease: hyperthyroidism can cause lid retraction, proptosis, and strabismus, treated by controlling thyroid disease, steroids for exacerbations, and strabismus or orbital surgery

Transient Visual Loss: called Amaurosis Fugax, can be due to carotid artery disease or migraine, pts need carotid artery evaluation (doppler ultrasound)

Myasthenia Gravis: can be systemic or isolated to the eye, causes muscle weakness, especially with repetitive use, can mimic any ocular nerve or motility problem, diagnosed with Tensilon test, can lead to respiratory weakness

SECTION 13: OCULOPLASTICS: THE ORBIT AND EYELIDS, A BRIEF REVIEW

The Basics

Orbit: the bony cavity in the skull that houses the eye and orbital tissues, made up of 7 bones and approximately 30 ml in volume

Openings in the Orbit: the orbital walls have various canals and fissures through which all veins, arteries and nerves to the eye and eye muscles, and several sensory nerves to face must pass

Lamina Papreciea: medial wall of the orbit, the thinnest wall, infection from the ethmoid sinuses can spread through the lamina papreciea to the orbit causing orbital cellulitis

Orbital Floor: inferior wall of the orbit, subject to fracture into the maxillary sinus with blunt trauma, most fractures need to be treated with antibiotics, steroids, and decongestants, can lead to orbital emphysema

Orbital Emphysema: air within the soft tissue of the orbit, comes from blowing the nose after an orbital fracture, all pts with orbital wall fractures need to be instructed NOT to blow their nose

Proptosis: one eye displaced forward more than the other (or both eyes forward), due to trauma, tumor, or infection within the orbit

Thyroid Orbitopathy: the number one cause of proptosis in adults

Orbital Cellulitis: infection of the tissues within the orbit (behind the orbital septum), the number one cause of proptosis in children

Preseptal Cellulitis: infection of the tissues of the eyelid (anterior to the orbital septum), can spread behind the septum causing orbital cellulitis

Orbital Septum: the connective tissue barrier that separates the preseptal tissues (eyelid) from the orbital tissues

CT Scan: good for looking at the bony walls and structures of the orbit

MRI Scan: good for looking at the soft tissues within the orbit

Orbital Tumors: in general, cause loss of vision from compression of the optic nerve (compressive optic neuropathy), and the tumors should be excised in most cases

Optic Nerve and Optic Nerve Sheath Tumors: also cause compressive optic neuropathy, however, excising these tumors will damage and kill the optic nerve, and the tumors may be observed

Enucleation: removal of the entire eye including sclera, usually the muscles are sewn to a donor sclera with an implant, performed for blind, painful eye

Evisceration: removal of the contents of the eye, leaving the sclera in place, the muscles are already attached to the sclera, so they are not moved, performed for blind, painful eye with endophthalmitis (the surgeon does not want the eye infection to spread to the rest of the obit by operating on the orbital contents)

Exenteration: removal of the entire eye and the orbital contents, performed for malignant orbital cancers

Retrobulbar Alcohol Injection: injection of absolute alcohol for blind painful eye (kills the pain nerves, at least temporarily)

Levator Palpebrae Superioris: muscle that lifts the eyelids, controlled by cranial nerve 3 (thus cranial nerve three palsy can lead to ptosis)

Orbicularis Oculi: muscle that closes the eyelids, controlled by cranial nerve 7

Tarsus: connective tissue that makes the eyelid firm

Canthus: Medial and Lateral canthus are the corners of the eyelids, give rise to canthal tendons that attach the eyelids to the bones of the orbit

Eyelid Structures: include skin, lashes, tarsus, grey line (muscle of Riolan which is part of the orbicularis oculi), conjunctiva, and meibomian gland orifices

Trigeminal Nerve: cranial nerve V1, that controls sensation to upper eyelid, eye, tip of the nose, and forehead

Maxillary Nerve: cranial nerve V2, that controls sensation to the lower eyelid, and cheek (maxillary region) of the face

Trichiasis: lashes touching the cornea in a variety of condition, can cause keratitis, corneal abrasion, red, irritated eye, and corneal ulceration and scarring

Assessing the Levator Palpebrae Superioris: check the pts Margin Reflex Distance (distance in mm of a light reflection from the central cornea to the superior lid margin), Levator Function (distance in mm the upper lid moves from downgaze to upgaze), and Palpebral Fissure (height in mm of the interpalpebral fissure with the pt looking straight ahead)

Signs of Eyelid Malignancies (Carcinomas): tumors of the eyelid that are malignant tend to cause ulceration, loss of the eyelashes, lid notching or distortion of the lid, and may bleed (benign tumors generally do not have these features)

- **Orbital Inflammatory Conditions**

Orbital Cellulitis: infection of the orbital tissues, pts have pain, proptosis, a red eye and limited motility, pts can lose the eye, need CT scan, treated with IV antibiotics in the hospital

Thyroid Orbitopathy: inflammation of the muscles and soft tissues of the orbit associated with thyroid disease, pts have proptosis, lid retraction, and strabismus, need CT scan, treated with thyroid control, steroids, strabismus surgery, and orbital decompression for extreme proptosis

Idiopathic Orbital Inflammation: called Orbital Pseudotumor, inflammation of the orbital tissues that mimics orbital cellulitis, autoimmune condition with pain, proptosis, and decreased vision, pts need CT scan and real orbital cellulitis or orbital tumor must be ruled out (or treated with antibiotics IV), then treatment is with systemic steroids (responds very rapidly)

- **Orbital Tumor Conditions**

Dermoid Cyst: congenital tumor of the eyelid, a mass under the skin in children, can be completely excised

Lymphangioma: congenital tumor of lymph channels, tumor can enlarge with viral infections, treated with limited excision only when necessary

Cavernous Hemangioma: number one primary benign orbital tumor of adults, treated with excision

Capillary Hemangioma: number one primary benign orbital tumor of children, can be treated with steroid injections or pill if causing amblyopia

Optic Nerve Glioma: tumor or the optic nerve, can be associated with neurofibromatosis, treated with observation or excision with enucleation or radiation

Optic Never Sheath Meningioma: tumor of the optic nerve sheath, treated with excision with enucleation for growing tumors

Rhabdomyosarcoma: number one primary malignant orbital tumor in children, causes rapid proptosis, may mimic orbital cellulitis, treated with biopsy, radiation, and chemotherapy

Orbital Lymphoma: white blood cell tumor of the orbit, may occur in any of the orbital tissues, pts need systemic lymphoma workup, treated with biopsy, radiation, and chemotherapy

Lacrimal Gland Tumors: can be due to lymphoma, infection (tuberculosis), or primary lacrimal cancer (benign or malignant), pts need CT scan, biopsy and treatment or excision depending on the etiology

Orbital Metastases: from neuroblastoma or leukemia in children, and due to breast or lung cancer in adults, treated with systemic chemotherapy to the specific cancer

Secondary Orbital Tumors: can spread to orbit from surrounding tissues of the sinus, eye, skin, conjunctiva, or brain

- ## Orbital Traumatic Conditions

Medial Orbital Wall Fracture: fracture of lamina papreciea, treated with antibiotics, decongestants, steroids, and observation, usually do not have to be fixed

Orbital Floor Fracture: fracture of the inferior wall of the orbit, can lead to inferior rectus entrapment, strabismus, and permanent muscle damage, pts need CT scan to look for entrapment, treated initially with antibiotics, decongestants, and steroids, and later surgical repair if there is muscle entrapment or the fracture is larger, the sign of entrapment is the inability to move the eye upward, and pts complain of diplopia

Orbital Hemorrhage: aka Retrobulbar Hematoma, acute bleeding into the orbit due to trauma or after retrobulbar block for surgery, treated with immediate lateral canthotomy and cantholysis (incising and releasing the lateral eyelid) to relieve the pressure from the hemorrhage before it causes damage to the nerve (compressive optic neuropathy)

- ## Eyelid Conditions

Cryptophthalmos: congenitally absent or small disorganized eye

Distichiasis: congenital second row of eyelashes, can scratch cornea causing keratitis, treated with hyphrecation or surgery

Epicanthus: congenital extra skin at the medial canthus, can give the appearance of strabismus when the eyes are actually straight, treated with observation

Epiblepharon: congenital hypertrophy of the orbicularis oculi muscle, can lead to trichiasis and keratitis, treated with epiblepharon surgery to rotate the lids and lashes

Ectropion: the eyelid is turned out, usually due to age or scarring of the skin (other causes exist), causes severe dry eye and exposure keratitis, treated with eyelid surgery (usually involving tightening the lower eyelid, may involve skin grafting)

Entropion: the eyelid is turned in, usually due to age or scarring of the conjunctiva (other causes exist), causes trichiasis and keratitis, treated with eyelid surgery, may involve tightening the lower lid or conjunctival grafting (taken from the buccal mucosa)

Ptosis: a droopy or low upper eyelid, due to age or other muscle or nerve problem, treated with ptosis surgery

Congenital Ptosis: due to poor levator development, treated with frontalis sling surgery if pt is developing amblyopia

Myogenic Ptosis: due to myasthenia gravis or chronic progressive ophthalmoplegia, a muscle disease causes the ptosis, may be improved with surgery

Aponeurotic Ptosis: due to age as the levator muscle slips off the tarsus, treated with levator advancement surgery

Neurogenic Ptosis: due to Horner's syndrome, a nerve disease causes the ptosis, treated with Muller's muscle resection surgery

Traumatic Ptosis: due to traumatic displacement of the levator off of the tarsus, pts are observed for six months (the ptosis may improve or resolve), the levator advancement surgery if necessary

Eyelid Retraction: usually due to thyroid disease, causes dry eye and keratitis, may be treated with levator recession surgery

Chalazion: chronic inflammation of an eyelid gland, treated with warm compresses, may need surgical excision

Hordeolum: acute inflammation of an eyelid gland, treated with warm compresses, may become chronic and turn into a chalazion

Verruca Vulgaris: a wart on the skin, caused by a virus, may spontaneously regress or be excised surgically,

Actinic Keratosis: a white, sand-paper-like scaling of the skin, due to sun damage, is pre-malignant

Keratoacanthoma: explosive volcano-like crater on the skin, simulates squamous cell carcinoma, treated with complete surgical excision although benign

Molluscum Contagiosum: papules with a central dimple, due to a virus, contagious, treated with excision or curretage

Nevus: a brown, pigmented mole on the skin, may be observed or excised if growing or suspicious

Basal Cell Carcinoma: malignant tumor of the skin, locally invasive, may look like a pearl or an ulcer on the skin, treated with excision

Squamous Cell Carcinoma: malignant skin tumor, invades lymphatics and lymph nodes (check the lymph nodes on exam), treated with excision and radiation may be necessary

Sebaceous Cell Carcinoma: malignant tumor of eyelid sebaceous cells, very aggressive locally and via the bloodstream, may mimic chronic conjunctivitis, difficult to excise surgically, may require exenteration of the orbital contents, very deadly cancer

Malignant Melanoma: malignant tumor of the skin from the pigmented melanocytes, pigmented tumor, may spread to liver, treated with surgical excision and evaluation for systemic spread, chemotherapy if necessary, also very deadly

Benign Essential Blepharospasm: spasmic contractions of the orbicularis oculi (lots of forceful, involuntary blinking) without underlying tumor or cause, treated with local botulinum toxin injection

Facial Palsy: due to damage to the seventh cranial nerve, can cause facial droop and lagophthalmos (the lid does not close all the way), treated with gold weight placement in the upper lid to assist in lid closure

Eyelid Lacerations: due to trauma, treated with surgery to close the lid defect and antibiotics, if the laceration is medial to the punctum, the canaliculus is likely to be involved and a silicone stent will be placed for several weeks or months in the canalicular system

Congenital Nasolacrimal Duct Obstruction: common in newborns, over 90% resolve within one year, if does not resolve, treated with probing and irrigation to open the valve of Hasner (membrane obstruction flow in the nose)

Acquired Nasolacrimal Duct Obstruction: due to infection or trauma in adults, treated with DCR surgery (dacryocystorhinostomy) to open a new drainage pathway from the lacrimal sac to the nose

Canaliculitis: infection of the lacrimal canaliculus, due to Actinomyces, treated with warm compresses, penicillin, and surgery if necessary to remove dacryoliths (stones that from in the canaliculus due to infection)

Dacryocystitis: infection of the lacrimal sac, due to many organisms, treated with oral antibiotics, incision and drainage if an abscess forms, and DCR surgery

• SECTION 14: PEDIATRIC OPHTHALMOLOGY AND STRABISMUS, A BRIEF REVIEW

• The Basics

Orthophoria- straight eyes, the eye muscles are keeping the eyes aligned
Strabimus (-tropia)- misalignment of the eyes (strabismus may be a cause of amblyopia, or longstanding amblyopia may lead to strabismus)
Strabismus in adults- if it is of new onset, often causes Diplopia (double vision) in adults
Strabismus in children- often does NOT cause Diplopia because the child will Suppress (turn off the input) from one eye. Suppression can lead to Amblyopia
Esotropia- eyes are misaligned inward
Exotropia- eyes are misaligned outward
Hypertropia or Hypotropia- the eyes are misaligned upward or downward
Phoria- any type of a Phoria (i.e. esophoria) is a tendency of the eye to turn, but the eyes are NOT misaligned
Nystagmus- rapid, involuntary eye movements
Pseudo-Strabismus- the appearance of strabismus when the eyes are actually straight (i.e. a child with a large nasal bridge or prominent epicanthal folds)
Extraocular Muscles- six muscles that move the eye, medial, inferior, lateral and superior rectus, and superior and inferior oblique, innervated by cranial nerves, 3, 4, and 6
Hering's Law of Equal Innervation: when the pt looks left, both the right medial rectus, and left lateral rectus receive equal and simultaneous innervation (the eyes move together because of Hering's Law)
Sherington's Law: contraction an an eye muscle is accompanied by the relaxation of the muscle's antagonist (with contraction of the right medial rectus, there is relaxation of the right lateral rectus)
Fusion: pts ability to place the image on the retina in the same place in both eyes
Diplopia: double vision, children with early onset strabismus often do not have diplopia because of suppression, adults with new-onset strabismus often have diplopia
Suppression: neural activity in children only that prevents the visual input from one eye from entering consciousness, can eventually lead to amblyopia
Amblyopia: a decrease in best corrected visual acuity in one eye due to strabismus (with suppression of one eye), visual deprivation (congenital cataract or high refractive error), or anisometropia (difference in refraction between the two eyes)
Amblyopia Treatment: first, pts need a cycloplegic refraction and to wear spectacles full-time, next the pts may need patching (placing a patch over the better seeing eye to increase the visual input from the amblyopic eye) or atropine penalization (one drop of atropine in the better seeing eye weekly to blur the image in the better eye and increase input from the amblyopic eye), patching and penalization is tapered and stopped as the vision improves
Strabismus Risk Factors: prematurity, family history, low birth weight and history of brain disease
Cover Test: have the pt fixate on a target, cover one eye, see if the other eye move to fixate on the target, then uncover both eye and repeat for the other eye, if one eye moves, the pt has a -tropia (manifest deviation)

Alternate Cover Test: have the pt fixate on a target, cove the right then left eye sequentially, if the eye move to fixate on the target the pt has a -phoria (latent deviation) or a -tropia (if the Cover Test was also positive)

Hirshberg Test: have the pt fixate on a light source, where the light reflex falls indicates an estimate of the prism diopters of strabismic deviation

Prism Diopter: the unit of measurement of strabismus, one prism diopter displaces light one centimeter at one meter's distance

Krimsky Test: a light is shown into the eyes and prisms are placed in front of one eye until the light reflexes are the same, this too estimates the strabismic deviation

Prism Cover Test: the pt fixates on a target, and an alternate cover test is performed while placing prisms in front of one eye, until there is no movement of the eyes, this measures the strabismic deviation

Maddox Rod Test: a maddox rod is placed in front of one eye, and the pt fixates on a light, if the light and line (from the maddox rod) are superimposed there is no -tropia or -phoria

Double Maddox Rod Test: a maddox rod is placed in front of each eye and the pt fixates on a light, the pt adjusts the rods until the lines are parallel, this tests for cyclodeviation

Cyclodeviation: a torsional diplopia, usually due to a disorder in the oblique muscles which account for much of the torsional movements of the eyes

Park's Three Step Test: alternate cover test in primary gaze (straight ahead), right and left gaze, and right and left head tilt, used for vertical deviations to determine the etiology

Dissociated Vertical Deviation (DVD): when one eye is covered (aka dissociated), that eye tends to drift upwards (vertically), and the same may happen for the opposite eye

Cycloplegic Refraction: any child with a strabismic deviation deserves a full-cycloplegic refraction with cyclopentolate for 30 minutes or atropine to identify and eliminate the refractive element of their deviation

Deviation: a -tropia or -phoria is a deviation from orthophoria, it is always measured relative to the other eye, often pts or family claim that one eye is turning but the other is straight, this may not be the case as the deviation is always measured relative to the other eye, this is important when explaining strabismus surgery that will involve both eyes, rather than just the one eye the pt or family perceives as turning

Head Position: a pt may maintain a certain head position to maintain fusion (keep the images of both eyes on the same part of the retina), such as chin-up in double elevator palsy or head-tilt in superior oblique palsy, head position is important because it helps diagnose the type of deviation and abnormal head position is one indication for strabismus surgery

- ## Pediatric Lid Conditions

Ptosis: can be congenital, associated with Horner's Syndrome, myasthenia gravis, or Marcus-Gunn Jaw Winking (see above), may be treated with ptosis surgery if amblyopia develops

Distichiasis: a congenital extra row of lashes, may cause keratitis, treated with hyphrecation or lid surgery

Trichiasis: a misdirection of the lashes rubbing on the cornea, can cause keratitis, treated with lid surgery

- ## Pediatric Infectious and Inflammatory Disorders

Ophthalmia Neonatorum: conjunctivitis in the newborn, caused by silver nitrate drops, or infection with Neisseria gonorrhoeae or Chlamydia, treated with systemic antibiotics along with topical drops (ceftriaxone for gonorrhoeae and erythromycin for chlamydia)
Preseptal Cellulitis: infection anterior to the orbital septum, does not threaten the the eye, but can lead to orbital cellulitis, pts have red, indurated eyelid, treated with oral or IV antibiotics and close observation
Orbital Cellulitis: infection of tissues posterior to the orbital septum, threatens vision and the eye, pt has pain on eye movement and red eye in addition to red, indurated eyelid, treated with hospital admission and IV antibiotics, commonly due to H. influenzae
Vernal Conjunctivitis: seasonal allergic conjunctivitis in boys, can lead to corneal ulcer and scarring, treated with topical mast-cell stabilizer, antihistamines, and steroids
Nasolacrimal Duct Obstruction: congenital blockage of tear drainage at Valve of Hassner, treated with massage, 90% resolve by 1 year of age, 10% treated surgically with probing and irrigation (probe is passed from the puncta to the nose to open the valve)

- ## Congenital Corneal Conditions

Microphthalmos: small, disorganized, poorly functioning eye
Nanophthalmos: small, but well-formed eye that functions
Buphthalmos: an enlarged eye (usually due to congenital glaucoma)
Microcornea: cornea less than 10mm
Megalocornea: cornea greater than 13mm
Sclerocornea: white cornea that looks much like the adjacent sclera
Angle Disorders: are problems with the formation of the anterior segment of the eye, including the angle, include posterior embryotoxon, and Axenfeld's and Rieger's anomalies, can be associated with glaucoma (because the angle is involved)
Peter's Anomaly: the lens is stuck to the cornea in front and the iris in back, presents as leukoma (white cornea), may be treated with surgery to replace the lens and cornea
Causes of White Cornea in Infants: there are many causes including birth trauma, congenital glaucoma, corneal infection, Peter's anomaly, congenital corneal problems, and Hurler's syndrome (a mucopolysaccharidosis)
Congenital Glaucoma: due to congenitally abnormal trabecular meshwork, can be associated angle disorders above, pts have tearing (epiphora), photophobia (light sensitivity), corneal edema, and buphthalmos, treated with IOP lowering drops (avoid brimonidine in infants), or surgery (typically goniotomy or trabeculotomy), trabeculoplasty and tube shunts if necessary
Iris Coloboma: congenital absence of part of the iris, causes a "keyhole" shaped pupil
Lens Subluxation: displacement of the lens inside the eye due to zonule weakness, associated with Marfan's syndrome and Homocystinuria
Unilateral Congenital Cataract: an opacity of the lens that is inherited, sporadic, or due to Peter's Anomaly, or PFV (fetal vessels that failed to regress during fetal

development)

Hereditary Bilateral Congenital Cataract: due to a systemic disorder such as Wilson's disease, Fabry's disease, Myotonic dystrophy, Down syndrome, Alport syndrome, Galactosemia, or Lowe syndrome

Infectious Bilateral Congenital Cataract: toxoplasmosis, rubella, CMV, herpes, or syphilis

Congenital Cataract Treatment: pts with bilateral cataracts need tests for the above infections, treated with cataract surgery with lens implant or contact lens use after surgery

• Pediatric Uveitis Conditions

Juvenile Rheumatoid Arthritis: autoimmune disease with arthritis, causes chronic eye inflammation, treated with topical steroid drops and systemic steroids and immunosuppressives

Intermediate Uveitis: inflammation of the pars plana, usually has no underlying cause, 25% of uveitis in children, treated with topical steroid drops, systemic steroids and immunosuppressives if necessary

Toxocariasis: infection due to Toxocara canis, spread by dog feces, causes retinal inflammation and detachment, treated with vitrectomy and retinal detachment surgery

• Pediatric Retina and Vitreous Conditions

Persistent Fetal Vasculature (PFV): fetal vessels that connect the lens to the optic nerve fail to regress during development, can lead to cataract, treated with cataract surgery

Retinopathy of Prematurity (ROP): occurs in premature, low birth weight infants, causes retinal neovascularization and retinal detachment, examination necessary for infants under 1500 grams or younger than 34 weeks gestational age at birth, classified by disease extent, dilated exam is performed every 2 weeks, treated with cryotherapy or laser if neovascularization occurs

Coat's Disease: abnomally formed retinal blood vessel causes retinal detachment, treated with laser to close the vessel, causes whitening of the retina that look like retinoblastoma on exam

Stickler Syndrome and Wagner Syndrome: two diseases with abnormal vitreous (called optically empty) and retina, subject to retinal breaks, treated with laser if necessary

Retinoschisis: a weakness of the retina splitting and separating the inner and outer retinal layers, associate with Goldmann-Favre disease and Juvenile X-Linked retinoschisis, treated with laser or surgery to peripheral retinal breaks as necessary

Retinitis Pigmentosa: inherited retinal disease that slowly causes night blindness, visual field loss, and ultimately complete blindness, vitamin A and acetazolamide may be used, no good treatment to prevent disease progression

Congenital Stationary Night Blindness: inherited retinal disease, causes night blindness, does not progress with time, no treatment necessary

Stargardt's Disease: inherited disease causing loss of central vision at an early age

Best's Disease: causes loss of central vision at an early age, has an abnormal electrooculogram (the one disease that is definitively diagnosed with an abnormal EOG)

- **Pediatric Optic Nerve Conditions**

Optic Disc Coloboma: part of the optic disc and retina are missing, associated with several systemic disorder, the heart may be affected and pts need systemic and cardiac workup

Optic Nerve Pit: common congenital defect in optic nerve formation, looks like an excavation of part of the optic disc, no associated systemic problems, can leak fluid leading to retinal detachment

Morning Glory Optic Nerve: congenital nerve defect, looks like a "morning glory" flower, causes decreased vision, pts need MRI to rule out brain abnormality, treatment with patching may improve vision some

Myelinated Nerve Fiber Layer: white appearing areas on the retina, where myelin extends past the optic nerve head (usually myelin is only present in the optic nerve when it is outside of the eye)

Optic Nerve Hypoplasia: congenitally small optic nerve, causes decreased vision, associated with brain abnormalities (Septooptic Dysplasia), pts need MRI and testing for endocrine (pituitary) abnormalities

Optic Nerve Atrophy: death of the optic nerve, can be due to infection, tumor, or hereditary disease, pts need MRI of brain and orbits to look for tumor or brain pathology, and Syphilis testing

Leber's Hereditary Optic Atrophy: inherited progressive optic neuropathy that causes severe loss of vision in both eyes around age 15 to 30, pts should avoid alcohol and tobacco

- **Pediatric Tumor Conditions**

Preseptal and Orbital Cellulitis: may masquerade as tumors, children with these conditions need a CT scan (or MRI) to image the orbit and see the extent of the infection

Orbital Pseudotumor: autoimmune inflammation of the orbital tissues, causes proptosis and pain, pts need CT scan imaging, treated with oral steroids

Capillary Hemangioma: benign blood vessel tumor of the orbit and eyelid, regresses spontaneously after one year, may be treated with oral steroids or steroid injection if large and causing amblyopia

Lymphangioma: benign lymph tissue tumor, causes proptosis that may worsen with viral illness, surgery may be necessary if hemorrhage occurs into the tumor causing it to enlarge rapidly and compress the optic nerve

Rhabdomyosarcoma: number one malignant orbital tumor of children, causes rapid proptosis, pts need CT or MRI, biopsy, and radiation or chemotherapy to treat the cancer

Neuroblastoma: number one metastatic orbital tumor or children, treated with radiation and chemotherapy

Optic Nerve Glioma: number on tumor or the optic nerve in children, causes proptosis and loss of vision, pts need CT or MRI, may be serially observed or treated with radiation of surgical excision with enucleation

Optic Nerve Sheath Meningioma: less common in children, may be observed or treated surgically

Retinoblastoma: a malignant eye tumor, the number one intraocular tumor of children, may be bilateral, pts have decreased vision or leukocoria (white pupil), need CT or MRI, can have associated brain tumor, treated with enucleation, laser, radiation, and chemotherapy if necessary, these children need frequent follow up exams under anesthesia to check the second eye and the side of the enucleation for any recurrence of tumor

Dermoid Cysts: benign cyst forming at superotemporal orbit, causes a mass, may be excised surgically

The Phacomatoses: systemic inherited diseases with tumors in multiple parts of the body, may be associated with skin findings (cafe au lait) spots, skin nodules and tumors, mental retardation, and brain tumors, include tuberous sclerosis, ataxia-telangiectasia, Sturge-Weber syndrome, Wyburn-Mason syndrome, and Neurofibromatosis types I and II

- ## Craniofacial Malformation Conditions

Goldenhar Syndrome: congenital abnormality with dermoid tumors on the eyes and preauricular appendages (abnormal tissue in front of the ears), teated with surgery for dermoids if causing astigmatism and amblyopia

Craniosynostosis Syndromes (Crouzon's Disease, Apert Syndrome, Pfeiffer Syndrome): the skull sutures close to early and fuse, causing proptosis and strabismus, pts have other cranial abnormalities

Fetal Alcohol Syndrome: due to fetal alcohol exposure, causes telecanthus (eyes far apart), ptosis, strabismus, and optic nerve hypoplasia, prevented by pregnant mothers avoiding alcohol

- ## Inherited Ocular Conditions

Albinism: lack of melanin pigment in the eyes or the eyes and skin, causes decreased vision and eye and skin depigmentation, pts may have severe decreased vision with nystagmus, can be associated with bleeding or blood cell disorders, treated with sunlight avoidance and UV protection for skin and in spectacles

Down Syndrome: trisomy 21, more common in children with mother over 35 years old, pts have mental retardation and high incidence of prominent epicanthal folds, strabismus, cataract, and Brushfield spots (small tumors on the iris), pts need good refraction and follow-up and surgery as necessary

Aicardi Syndrome: optic nerve hypoplasia and retinal disorder, seen only in girls, fatal in boys

- ## Nystagmus Conditions

Congenital Nystagmus: caused by sensory deprivation (no visual input into both eyes), can be due to aniridia, bilateral congenital cataracts, retinitis pigmentosa, ocular albinism, optic nerve hypoplasia and other causes, pts have involuntary rhythmic movements of the eyes, develops at 3 months of age, occasionally treated with strabismus surgery to lessen the amplitude (decrease the size) of the nystagmus movements

Latent Nystagmus: nystagmus in pts with congenital esotropia that only appears when one eye is covered

Spasmus Nutans: high frequency rhythmic eye movement with head nodding, pts to CT or MRI to rule out tumor, may resolve spontaneously

- ## Strabismus Conditions

Congenital Esotropia: eyes turn in, also called infantile esotropia because onset from 0 to 6 months of age, associated with DVD, treated with spectacles and patching for amblyopia first, then strabismus surgery (giving the pt the full cycloplegic refraction is essential to remove any accommodative component of the esotropia)

Accommodative Esotropia: aka Refractive Esotropia, in young pts with high plus refractive error, pts must accommodate to focus the eyes and the acts of accommodation and convergence are linked, therefore pts become esotropic because of the accommodation and convergence effort, treated with cycloplegic, high plus spectacles and may be treated with bifocals

Exotropia: eyes turn out, can be due to mechanical problems (thyroid eye disease), poor convergence (congenital), or congenital brain disease, treatment includes minus lens spectacles (to stimulate accommodation and convergence) and surgery

Intermittent -tropia: an eye deviation that is not constant, pts must be followed closely for worsening to a constant -tropia

Double Elevator Palsy: the eye is turned down and will not elevate (go up) on testing, may cause a chin-up head position to maintain fusion, treated with strabismus surgery

Superior Oblique Palsy: the eye is turned up, congenital or due to head trauma, causes head-tilt position to maintain fusion, diagnosed with Park's three step test, treated with strabismus surgery

Brown Syndrome: eye cannot elevate in adduction (gaze toward the nose), due to superior oblique tendon problem, may cause chin-up head position, treated with steroid injection or strabismus surgery

A Pattern Deviation: any -tropia that is measured larger in downgaze, like an "A" is larger at the bottom of the letter, associated with overaction of the superior oblique

V Pattern Deviation: any -tropia that is measured larger in upgaze, like a "V" is larger at the top of the letter, associated with overaction of the inferior oblique

Duane Retraction Syndrome: eye cannot abduct (turn toward the temple) and the eye retracts backward into the orbit (looks smaller) on adduction (turning toward the nose), may cause face turn head position, associated with Goldenhar syndrome, treated with strabismus surgery

Moebius Syndrome: multiple cranial nerve palsies with strabismus and facial weakness

Thyroid Eye Disease: number one cause of strabismus in adults, treatment includes steroids for acute exacerbations of inflammation, prism glasses for diplopia, and orbital decompression surgery or strabismus surgery

• REFERENCES AND RECOMMENDED READING

ASORN. (American Society of Ophthalmic Registered Nurses). **Ophthalmic Procedures: A Nursing Perspective Office & Clinic**. Kendall/Hunt, Dubuque, 2006. Available at http://kendallhunt.com

Boyd-Monk, H, Steinmetz, C. **Nursing Care of the Eye**. Appleton & Lange, Norwalk, 1987.

Chern, K, Zegans, M. **Ophthalmology Review Manual**. Lippincott, Williams & Wilkins, Philadelphia, 2000.

Goldberg, S. **Ophthalmology Made Ridiculously Simple**. MedMaster, Miami, 2005

Goldblum, K, Lamb, P (Eds.). **Core Curriculum for Opthalmic Nursing**. Kendall/Hunt, Dubuque, 2006. Available at http://kendallhunt.com

Kunimoto, DY, Kanitkar, KD, Maker, M, Friedburge, MA, Rapuano, CJ. **The Wills Manual: Office and Emergency Room Diagnosis and Treatment of Eye Diseases**. Lippincott, Williams & Wilkins, Philadelphia, 4th ed., 2004

Pavan-Langsten, D. **Manual of Ocular Diagnosis and Therapy**. Lippincott, Williams & Wilkins, Philadelphia, 5th ed. 2002.

Physician's Desk Reference for Ophthalmic Medicines. Thompson Healthcare, Inc. New Jersey, current edition.

Tasman, W, Jaeger, E. **Duane's Clinical Ophthalmology**. Vol. I-VI, Lippincott, Williams & Wilkins, Philadelphia, current edition.